HEAT AND LIFE

HEAT AND LIFE

The Development of the
Theory of Animal Heat

EVERETT MENDELSOHN

HARVARD UNIVERSITY PRESS

Cambridge, Massachusetts

1964

TO MARY

Preface

Historians of science have paid much attention in recent years to those problems which might be described as indicating a "watershed" or "scientific revolution." The attempt has been made to map the period of great change, and where possible to delineate the structure or morphology of the change. This new approach to the study of the scientific past has reflected the outlook of the historian of ideas—albeit this time the study is of scientific ideas. The areas of greatest success—astronomy, mathematical physics, and other parts of the exact sciences—were those in which the concepts under scrutiny could be manipulated independently of a large mass of experimental or experiential data. The shifts in cosmology or changes in the explanation of motion were shown to be alterations primarily of ideas; often the same data served to support both the old and the new analysis. So frequently have these areas of the physical sciences been called upon as the examples in the study of scientific change that we might almost conclude that they suffice as the models for all of science. In an attempt partially to rectify this situation I have turned to a study in the history of biology.

The biological sciences have seldom produced concepts that were as independent of observation and experience as were those of the physical sciences. Where this was the case, with a concept such as *life*, it was so diffuse as to be of little help in directing further scientific investigation. Most theories in biology, especially functional biology, have been closely tied to a single organ, a specific function, or a special group of living things. It is no accident that it was biology, among all the sciences, that most tried the patience of René Descartes when he attempted valiantly, if unsuccessfully, to reduce the phenomena of life to generalized physical laws. The use of reduction as a simplifying device in establishing a new theory in biology is found throughout the modern period. At the same time it has often been severely criticized by those who viewed such a step as philosophically unacceptable when dealing with living things.

The question that the biologist often wants answered is whether this reduction of a biological phenomenon to a physical or mechanical analogue has been helpful in building a new theory, or whether one must expect to find strictly biological theories to explain biological phenomena. The only way to find the answer is to turn to the history of biology and examine in detail a theory or concept that has been through the rigors of reduction and attempt to discover where and under what circumstances the novelty or innovation really arose.

There are further questions that the student of biological thought wants answered. He wants to know what relation exists between the new biological theory and the physical model from which it received its inspiration. Was it new or old physical theory, or physical technique, to which the biologist had recourse? Was it indeed a biologist who proposed the new physically oriented biological theory? Did the biological problem serve in any way to encourage new research and theory building in the specific area of the physical sciences? It is obvious that the biological theory chosen for scrutiny makes quite a difference to the range of questions that can be answered.

Animal heat provided me with just the right type of example. In one form or another the heat of the warm-blooded animal was recognized and explained from the earliest times. The obvious necessity of reconciling biological and physical heat was apparent at all periods. Perhaps even more important was the seeming accessibility of body heat for sensory observation. What physician had not noticed fevers and chills in the course of the study of illness?

This study, then, had its origins in my desire to find out what elements went into concept formation in the biological sciences and what the relation of this process was to the knowledge and techniques of the physical sciences. A single theory, and the changes it has undergone, is here examined in detail. The interplay of the new and the old, of thought and technique, is everywhere apparent. The elements to which we now attribute "modernity" can be traced in their emergent forms. I have not attempted to carry the historical narrative to the present, primarily because the pattern established in the decade just before the nineteenth century began was the critical new element. Furthermore, the scope of the problem broadens to such an extent during the nineteenth century that we can no longer view the changes in a single theory, but become involved in the mutual influences of a number of interacting theories. And this provides the basis for a study in which I am now engaged.

Research seldom involves just one person or even an easily recognizable group of people. The debts owed to colleagues include direct assistance as well as indirect inspiration. Clearly it is impossible to give just acknowledgment to all. Several words, however, must not be omitted. Much of the work behind this book was carried out while I was a Junior Fellow of the Society of Fellows at Harvard University. The encouragement to intellectual exploration received while a member of that scholarly consortium will keep me in their debt throughout my professional life. I. Bernard Cohen, Henry Guerlac, and Erwin Hiebert were each responsible for many helpful suggestions during the course of writing and revision of this book. That I have benefited from their advice and criticism will be apparent to each. That I have not taken all their words to heart leaves the faults of the book mine alone. Librarians at the Harvard College Library and the University Library, Cambridge, England were helpful in securing books and answering puzzling queries.

I am thankful to the editors and publishers of the *Mélanges Alexandre Koyré: Aventure de la Science* (Paris: Hermann, in press) for permission to include a revised version of my contribution, "The Nature of Physiological Explanation in the Seventeenth Century," as part of Chapter III of this work.

In an earlier stage, this material was presented to the Committee on the History of Science, Harvard University, as my doctoral dissertation.

The secretarial assistance received during the course of writing and rewriting a book is often of a value much greater than that implied in the job description. Miss Betsy A. C. Smith and Mrs. Timothy Little both gave assistance in manuscript preparation. The completion of this study and the accuracy of citation owe much to the careful attentions of Mrs. Philip Perlman.

Final appreciation goes to family and friends who gave generous encouragement at each step.

E. M.

Contents

HEAT AND LIFE

HEAT AND LIFE

· I ·

Introduction

"Life," wrote Claude Bernard, "is nothing but a word which means ignorance, and when we characterize a phenomenon as vital, it amounts to saying that we do not know its immediate cause or its conditions."[1] The whole history of biological thought can be outlined as a transition from a period when men were willing to attribute a living phenomenon to a "vital force" to that stage when they demanded a description of the conditions and a search for the material causes of all living activities. The difference in the theories and concepts developed at these two levels of thinking in biology is profound, although on the surface the biologist's description of any single phenomenon may be almost the same at all times.[2]

In forming explanations of the activities of living systems the biologist has often had recourse to comparison and analogy. This is probably due to the fact that most biologi-

[1] Claude Bernard, *An Introduction to the Study of Experimental Medicine*, trans. H. C. Greene (New York: Schuman, 1949), p. 201.

[2] Claude Bernard, for example, in his *Leçons sur la Chaleur Animale, sur les Effets de la Chaleur et sur la Fièvre* (Paris, 1876), p. vi, commented "that heat is a condition essential to the manifestation of the phenomena of life. The heat which has its source in the interior of the organism, in man and warm-blooded animals, explains how their vital activities remain free and independent, within limits, of external climatic variations." This descriptive statement would have been as acceptable to the ancients as it is to the modern physiologist.

cal "explanation" is in large measure descriptive in nature, and description is fundamentally comparative.[3] Analogies, however, involve more than comparison between gross phenomena; they have often been used in attempts to understand the relations existing between activities in one process by seeking a resemblance to already explained relations in the analogous process. Thus, in Greek antiquity explanations of the physical universe were cast in terms more naturally applied to the living organism. Medieval natural philosophers, on the other hand, having inherited the doctrine of the four elements and four humors, felt it completely reasonable to suppose that there were four organs or parts of the body to correspond.[4] The similarity seen between animal heat and combustion, in the seventeenth century, certainly serves as an example of the type of analogy used in early modern science.

There was, however, more than a change in the subject matter of analogies: there was a general shift in the use to which analogy was put in science. In the Middle Ages conclusions drawn from analogy were deemed to need no further proof.[5] Modern science came to utilize analogy primarily as a source of suggestions for hypotheses, especially in the early stages of investigation of a subject. But even when analogy is used just as a guide, with independent proof being sought, the scientist, whose thought is in great measure molded by the thought patterns of his age, is unaware of the extent of his indebtedness to analogy. In biol-

[3] See Agnes Arber, "Analogy in the History of Science," in *Studies and Essays in the History of Science and Learning,* ed. M. F. Ashley Montagu (New York: Schuman, 1944), p. 223; also Michael Scriven, "Explanation and Prediction in Evolutionary Theory," *Science,* 130 (1959), 477–482.

[4] Charles Singer, "A Review of the Medical Literature of the Dark Ages, with a New Text of About 1110," *Proc. Roy. Soc. Med.* (Sect. Hist. Med.), 10, part 2 (1916, 1917), 123.

[5] Arber, "Analogy in the History of Science," p. 229.

ogy a great deal can be learned about the stage of development of a given study by the choice of analogies used to illustrate a problem.

The history of biology gives evidence of the growing reliance of the biologist on mechanical and physical analogy. But, contrary to the common notion, the use of analogies drawn from the nonorganic world does not of necessity indicate that the user is giving a mechanistic interpretation of biological processes. Aristotle (hardly to be considered a mechanist), when talking at one point about the formation of parts of the body, commented by comparison that "exactly the same happens with things formed by the processes of the arts. Heat and cold soften and harden the iron, but they do not produce the sword."[6] In his discussion of the "vital flame" Galen compared it to the fire in a lantern, finding the fat of the body playing the same role as the oil of the lamp, and respiration serving the function of ventilation in a manner similar to the stream of fresh air supplied to the flame in the lantern.[7] The flame in the heart, however, remained qualitatively different than the flame of common fire.

What is of particular interest is the manner in which physical and mechanical analogies are utilized in the genesis of biological concepts and theories. For Aristotle and Galen it is quite clear that the physical systems are described by way of providing simple illustrations for complex, and often hidden, organic phenomena. The living system, however, they believed to be governed by laws peculiar to itself and not shared by the inorganic world. By the seventeenth and eighteenth centuries the recourse to physical and me-

[6] Aristotle, *Generation of Animals*, trans. A. L. Peck (Loeb Classical Library; Cambridge, Mass.: Harvard University Press; London: Heinemann, 1943), 734 b 37–735 a 2.
[7] See Chapter II for a full discussion of Galen.

chanical analogy was often indicative of an attempt to fill
the gaps in an explanation of a biological process by utilizing
the steps found in an analogous nonliving system. Biological
phenomena, the modern authors implied, could be explained
by the same theories and concepts acceptable in explaining
the processes of the nonorganic world. By the end of the
eighteenth century animal heat, for example, was no longer
viewed as being "like" combustion; it was instead described
as a "slow form of combustion."

It is this change in the explanation given for the heat of
animals which Jacques Loeb identified as signifying the
beginning of scientific biology.[8] For Loeb the point is quite
simple: biology becomes scientific just as fast as it is able
to substitute physicochemical explanations for special biological explanations. The reduction of this life phenomenon
to physicochemical terms, Loeb thought, gave biologists
their first real entrance to the solution of the problem of
life.

As we survey, in the following pages, the step-by-step
alterations in a single physiological theory, several points
should emerge quite clearly. It will be noticed that the new
advances toward the establishment of a "scientific biological" explanation for animal heat were produced through
the application of the new chemistry and the new physics
of the day. Perhaps even more startling is the recognition
that in this area of biology it is often the men trained in
physics and chemistry who were responsible for the new
work.

We might almost say that beneath the concepts a man
develops, and underlying the type of experimentation he

---

[8] Jacques Loeb, *The Mechanistic Conception of Life* (Chicago, 1912),
pp. 4–5.

carries out, lies his orientation to the phenomena of nature. The life-oriented writers and experimenters developed life-oriented concepts. With the steady invasion of the fields of functional biology by men oriented, through training, to the sciences of the inorganic realm new approaches to biological activities emerged. The concepts these men formed reflected their desire to understand nature in terms of its physical and chemical components. Even when a "vital force" might be admitted as a "final" explanation, the techniques of experimentation had come to demand a knowledge and competence in the more exact sciences.

By the time Lavoisier, Crawford, and their contemporaries had finished their work in the last years of the eighteenth century this whole area of physiology had been so thoroughly invaded by physics and chemistry (as well as by the physicist and the chemist) that the physiologist emerged as a new man. Physiology had been transformed and a thorough understanding of physics and chemistry became a requisite for contributing to the advance of this area of biology.

Recognition of this transformation was not immediate, and some, like the anatomist-physiologist François Xavier Bichat, vigorously objected to the implications. While he agreed that "one calculates the return of a comet, the speed of a projectile," he argued that "to calculate with Borelli the strength of a muscle, with Keil the speed of blood, with Lavoisier the quantity of air entering the lung is to build on shifting sand an edifice solid itself but which soon falls for lack of an assured base."[9] The phenomena of life, he contended, do not exhibit the regularity and uniformity

[9] Xavier Bichat, *Physiological Researches on Life and Death*, trans. F. Gold (Boston, 1827), p. 78.

that we have come to expect in physical phenomena. Special vital laws and with them biological means of experimentation seemed necessary.

By the mid-nineteenth century, however, even for the physiologist who, like Rudolf Virchow, wished to leave room for the action of a vital force, there was "no disagreement regarding the fact that chemical and physical investigation is primary, and anatomical and morphological investigation secondary."[10]

The theory of animal heat offers a particularly good instance in which to examine the role played by new developments in the physical sciences in the formation of a biological theory. Heat was generally treated as a phenomenon of the physical world and yet its physiological analogue was recognized from earliest antiquity. The similarity of the heat found within and outside the animal body made comparison of the nature of the two types of heat inescapable. It also made this biological theory extremely susceptible to changes in the analogous physical theory. Thus, upon examining the theory of animal heat we notice its close reliance upon developments in the physics of heat and also recognize it falling into the same difficulties as suffered by the general theory of heat. The early inability to distinguish between thermal intensity and thermal capacity left its mark of confusion upon theories of vital heat. The controversy over the nature of heat—was it a substance, or merely the motion of parts?—is clearly reflected in the discussions of physiological heat. Following upon the successful discovery of the nature of combustion, the physiologist turned to the body, assuming that there

[10] Rudolf Virchow, "Cellular Pathology (1855)," trans. in Lelland J. Rather, *Disease, Life, and Man, Selected Essays by Rudolf Virchow* (Stanford: Stanford University Press, 1958), pp. 84–85.

would be found therein a form of combustion compatible with the living system. Biology was not the only beneficiary; a good case can be made to show that the necessity of finding a theory of heat which would be equally applicable to the living and nonliving worlds served to stimulate research in areas of heat theory that might otherwise have been ignored.

Although the ability to deal with heat was the major limiting factor throughout the development of the theory of animal heat, hindsight has shown one or two other questions which, had they been solved, would have greatly aided the construction of a consistent theory. Important among these is the respiratory process, which was recognized almost from the beginning as being closely tied to the heat of the body, and yet which defied precise solution into the nineteenth century. The simple inability to deal with the respiratory gases made it impossible to discern the full function of this important process. The reliance of the theory of animal heat on conceptual and experimental advances in physics and chemistry forced upon this physiological theory a rate of growth closely limited by developments in the nonbiological sciences.

·II·

"Innate" Heat

The theory of an "innate" body heat was first proposed by the early Greek medical and philosophical writers. As developed by them, this innate heat became the single most important motive power in the animal system. It was responsible for the generative and growth functions; it played a major role in effecting digestion; it was necessary for movement, sensation, and thought. The maintenance of innate heat coincided with life, the destruction of innate heat with death.

The strength of the theory lay in its postulating a source of heat and activity inherent in the living animal body, and indistinguishable from life itself. No cause for such heat need be sought, nor indeed was a cause deemed necessary by most of those who utilized the theory. The effects could be examined and the conditions requisite for the continuance of the innate heat explored without questioning the origin of this arbiter of animal life.

It is impossible to decide why the theory of innate heat was first proposed. But it is clear that a generalization as powerful as the one contained in this theory made possible the unification and explanation of otherwise very diverse phenomena. It accounted for the appearance of heat when there was no obvious source of heat. The sperm, for example, was believed to be cold, yet the developed organism was clearly warm.

In Plato's *Timaeus* the "hot" (one of the traditional opposite powers) is identical with the vital heat.[1] Plato attributed to the vital heat, or fire, a cutting power which enables it to aid in the digestion of food and drink.[2] The principal seat of the internal fire is the heart, which consequently must be continually cooled. According to Plato the lung was placed "around the heart as a sort of buffer, so that, when the spirit therein was at the height of passion, the heart might leap against a yielding substance and be cooled down."[3] Respiration, then, served for Plato to moderate the innate heat.

A similar theory, regarding the site of the innate heat and the refrigerating effects of respiration, was held by the author of the *Peri kardies*, or treatise "On the Heart." The "inborn fire" is located in the left ventricle of the heart, for which reason the ventricle is "thickly constructed within, in order to preserve the strength of the heat." The right ventricle is considered cool, and the blood which may seem warm "is not in its nature hot" although like any other "water" it can take up heat. The heart is comfortably enwrapped in the lungs, which are naturally cool and also are cooled by inhalation.[4] Thus, the doctrine of internal heat was a decisive factor in the cardiac physiology of both Plato and the author of the "Hippocratic" treatise. The

[1] Friedrich Solmsen, "The Vital Heat, the Inborn Pneuma, and the Aether" *J. Hell. Stud.*, 77 (1957), 119.

[2] F. M. Cornford, *Plato's Cosmology, The 'Timaeus' Translated with Commentary* (New York, 1937), p. 314.

[3] *Ibid.*, pp. 283–284, *Timaeus* 70 C-D.

[4] Frank R. Hurlbutt, Jr., "*Peri kardies*. A Treatise on the Heart from the Hippocratic Corpus: Introduction and Translation," *Bull. Hist. Med.*, 7 (1939), 1111, 1113, *Peri kardies* V, VI, XII.

action of the heart, the structure of the heart, and the function of respiration were all designed to accommodate the innate heat.[5]

Plato has adopted an explanation of respiration which is very similar to that attributed to Empedocles.[6] Respiration is treated in a mechanical fashion with the motive force being supplied by the "natural movement of the internal fire towards its own kind, and by the circular thrust so imparted to the air." The body breathes through both the mouth and pores in the skin; an inhalation at the mouth would coincide with an exhalation through the pores. Breath and blood move through the same channels; respiration and circulation are a single process driven by the innate heat.[7]

An important observation is made by Solmsen, who points out that Plato deliberately reduced the importance of the vital heat.[8] Although Plato, in the physiology of the *Timaeus*, definitely utilized the innate heat, he treated it on the same level as respiration and nutrition: a condition necessary to the functioning of the organism, but not a determining influence. Plato restricted the influence the vital heat had on life and death, on growth and decline. He made sure that the sphere of the *psyche* was not interfered with.

[5] Hurlbutt, "*Peri kardies*, Introduction," p. 1107, suggests that correct anatomical observations were reinterpreted to fit the physiological doctrines; for example, in the case of the pericardial fluid the author developed the theory that the fluid is augmented by swallowed liquids that seep through the epiglottis, the purpose being to prevent the heart from overheating. Hurlbutt's view might aid in understanding Plato's notion that the lung receives "breath *and drink*" (*Timaeus*, 70 C).

[6] The Empedoclean theory is outlined in the famous fragment 100. See John Burnet, *Early Greek Philosophy* (4th ed.; New York, 1930), pp. 219–220.

[7] Cornford, *Plato's Cosmology*, pp. 304, 306, 307; *Timaeus* 79 A-E.

[8] Solmsen, "The Vital Heat," p. 123.

Aristotle was not quite as cautious, in his use of innate heat, as was his master in the Academy. For Aristotle "the innate *heat* of the heart . . . is treated as the source of life and of all its powers—of nutrition, of sensation, of movement, and of thought."[9]

The nature of the vital heat is not always clear in Aristotle's works. In one section of the De generatione animalium he clearly states "that the heat which is in animals is not fire and does not get its origin or principle from fire."[10] The distinction between the vital heat and fire becomes blurred in other passages, especially those in which Aristotle draws an analogy between the internal warmth and a fire.[11]

Although all parts of the animal body possess some warmth, the source of the heat is in the heart in sanguineous animals and in the corresponding organ of other animals. It is for this reason that Aristotle believed that an animal might live when other parts of the body are cooled but would die as soon as the heart lost its heat. Not only is the heart the center of warmth but it is also there that "the soul is, as it were, set aglow with fire."[12] Some, said Aristotle, believe the soul of the animal to be identical with fire, but it would be better to say that "the Soul subsists in some body of a fiery nature." This, because heat is the most serviceable for the soul's activities, to nourish and to cause

[9] Sir David Ross, *Aristotle, Parva Naturalia, A Revised Text with Introduction and Commentary* (Oxford: Clarendon Press, 1955), p. 42.

[10] Aristotle, *De generatione animalium*, 737 a 7–8; *Generation of Animals*, trans. A. L. Peck (Loeb Classical Library; Cambridge, Mass.: Harvard University Press; London: Heinemann, 1943).

[11] *De iuventute*, 469 b 15, ed. W. D. Ross (London, New York: Oxford University Press, 1942). *De partibus animalium*, 652 b 7; *Parts of Animals*, trans. A. L. Peck (Loeb Classical Library; Cambridge, Mass.: Harvard University Press; London: Heinemann, 1955).

[12] *De iuventute*, 469 b 5–8, 15–16. See also *De partibus animalium*, 170 a 23, where the heart is likened to a hearth.

motion.[13] In locating the vital heat in the heart, Aristotle has given to this organ a preeminence in body functions, and designed much of his physiology to cater to its needs.

Nature has constructed both the lung-breathing and the gill-using animals in a manner consonant with the preservation of the innate heat.[14] The fire in the heart is preserved by the refrigerating effect of respiration. For, as Aristotle noted in a comparison of the vital heat to flame, a fire can be put out not only by extinction, but also by exhaustion, from its own excess. "Clearly therefore, if the bodily heat must be conserved (as is necessary if life is to continue), there must be some way of cooling the heat resident in the source of warmth."[15]

The preservation of the internal warmth through breathing does not in any way involve the adding of heat. "It is . . . nonsense that respiration should consist in the entrance of heat, for evidence is to the contrary effect; what is breathed out is hot, and what is breathed in is cold." In these passages Aristotle seems to be combating a Hippocratic notion that respiration exists in part for nourishing the internal fire. How, Aristotle asks, "are we to describe this fictitious process of the generation of heat from the breath?"[16]

But respiration is not alone in its service as a preserver of the vital heat. Aristotle believed that nature had also contrived the brain as a counterbalance to the heart and the heat it contains. Following this mode of thought Aristotle concluded that "although all blooded animals have a brain, practically none of the others has . . . , for since they lack

---

[13] De partibus animalium, 652 b 7–11.

[14] De somniis, 456 a 8–11. See also De partibus animalium, 668 b 33.

[15] De iuventute, 469 b 21f, 470 a 5 .

[16] De respiratione, 472 b 33–5, 473 a 10–11. See Ross, Aristotle, Parva Naturalia, p. 312.

blood they have but little heat."[17] Furthermore, the large size of the brain of man is due to the great amount of heat and blood in the heart.

The innate heat controls growth through the concoction and transformation of the fluid and solid matter of food.[18] The process of digestion proceeds from the gut into the blood, with the internal warmth at all times active. Aristotle has substituted a cooking of the food for the cutting described by Plato.[19]

The mode of animal reproduction is also determined by the vital heat. The warmer an animal is, the more perfect will be the state in which its young are generated. Live young are produced by the hotter animals; colder ones produce eggs; the coldest of all, such as insects, produce a larva which in turn produces an egg.[20] And as the vital heat presides over generation, so too over death. Violent death is due to the rapid extinction or exhaustion of the innate heat, while natural death is the exhaustion of the heat due to lapse of time and insufficient cooling.[21]

Thus Aristotle has found in the innate heat an instrument through which he can explain the myriad functions of the living organism, from its very generation to its passing away. What is the nature of the vital heat? It is like fire, but not quite like fire; it needs fuel,[22] but even more it needs to be moderated and cooled to keep it from burning to exhaustion. Aristotle's views on the innate heat are im-

[17] *De partibus animalium*, 652 b 24–6.
[18] *Ibid.*, 650 a 1–5. Also *De sensu*, 442 a 4.
[19] Solmsen, "The Vital Heat," p. 119.
[20] *De generatione animalium*, 733 a 34–b 17.
[21] *De respiratione*, 474 b 13–24.
[22] In only one place was I able to find a direct reference to a source of the innate heat. *De respiratione*, 473 a 10: "Further, how are we to describe this fictitious process of the generation of heat from the breath? Observation shows rather that it is a product of the food."

portant, for together with the later contributions of Galen they formed part of the doctrine of the strongest and longest-lived biological and medical tradition that man has known. Well into the seventeenth century the vital heat, as defined by Aristotle, was utilized in explanations by the foremost physicians and physiologists. But what in Aristotle's hands had been a powerful generalization useful for the construction of homogeneous physiological laws became in later centuries a rigid doctrine used to avoid the finding of limited causal explanations in the living organism.[23]

Criticism of the pervasive power of the vital heat of the animal body did not have to wait until the seventeenth century. In at least two Greek works there is some disparagement of the role of heat in the living system. The author of the treatise *On Ancient Medicine*, from the Hippocratic corpus, is opposed to the type of generalization which proposes heat as the principle of nature, and the cause of sickness and health.[24] "For it is not heat which has the great power, but astringency and insipidity and the other things I mentioned, both in man and external to him." As a matter of fact our unknown author considers heat and cold the least potent of the powers of the body. This weakness he believes is due to the constant alternation

[23] The doctrine of the *pneuma*, though intimately connected with the innate heat, has been consciously bypassed in this chapter. I felt that the discussion necessary for making the *pneuma* understandable would extend the work considerably without casting any new light on the nature of the innate heat. The *pneuma* is discussed at some length in L. G. Wilson, "Erasistratus, Galen, and the *Pneuma*," *Bull. Hist. Med.*, 33 (1959), 293–314, and O. Temkin, "On Galen's Pneumatology," *Gesnerus*, 8 (1951), 180–189. See also Ross, *Aristotle, Parva Naturalia, passim,* and Solmsen, "The Vital Heat."

[24] For critical editions of the text see W. H. S. Jones, *Philosophy and Medicine in Ancient Greece, Supplements to the Bulletin of the History of Medicine* [*Ancient Medicine*], No. 8 (Baltimore: Johns Hopkins Press, 1946), and A.-J. Festugière, *Hippocrate, "L'ancienne médecine," Introduction, traduction et commentaire* (Paris: Klincksieck, 1948).

between hot and cold taking place in the body, for through blending the "cold moderates the heat, and heat moderates the cold." After all, asks this practical-minded physician, how do the believers in the postulate about hot and cold treat their patients? "For they have not discovered, I suppose, an absolute hot or cold or dry or moist participating in no other form. But they have to hand . . . the same foods and drinks as we all use, adding to one the quality of being hot, to another of being cold . . . since it would be futile to order an invalid to take 'some heat.' "[25]

Ancient Medicine was probably written in the very late fifth century b.c.,[26] that is, some time after the promulgation by Empedocles of the doctrine of opposite powers, and before a fully developed theory of innate heat was utilized by Aristotle. The work is clearly not in sympathy with the views of Philistion, Plato, or many of the Hippocratics, for all of whom heat had assumed a major role in the functioning of the living organism.

Praxagoras of Cos was willing to agree that heat had an important function in the animal, especially in the conversion of nourishment into blood, but he denied that warmth was inherent to the body; there was no innate heat. Even though heat is not considered an independent entity within the body, its strength still has to be "according to nature"; different amounts of heat are required by different animals for the changing of food into blood. If the amount of heat does not suit the nature and constitution of the animal, digestion will be abnormal.[27]

[25] Jones, *Ancient Medicine*, p. 78–79. G. E. L. Owen (Corpus Christi, Oxford) suggested to me that the author of *Ancient Medicine* was criticizing the doctrine of Philistion when writing this section.

[26] Jones, *Ancient Medicine*, p. 47.

[27] Fritz Steckerl, *The Fragments of Praxagoras of Cos and His School, Collected, Edited and Translated (Philosophia Antiqua*, vol. 8; Leiden: Brill, 1958), pp. 11, 36; see also p. 58. Steckerl traces an intellectual relation between the author of *Ancient Medicine* and Praxagoras, pp. 41–43.

The fragments[28] provide no indication of why Praxagoras denied that heat was inherent to the body. His most
recent editor concludes that he was "a rather earthy personality, a hard-boiled materialist who took his explanatory
notions mainly from the sphere of the senses." That
Praxagoras' distrust of generalizations extended to more
than innate heat is seen in his denial of the existence of the
equally sacred "innate pneuma." Perhaps it was the interest
of Praxagoras and his followers in food and its composition
that led them to doubt the inherent quality of heat. They
seemed to believe that what happens in the body is a direct
consequence of what is contained in the food.[29]

Although there is no direct answer to the question of how
heat is acquired by the body, it can be inferred from the
beliefs about food. Heat was probably thought to have
come into the body with nutriment, being conveyed either
by the temperature of the food itself or by a certain quality
or humor, hidden within the food, with which heat is
connected.[30]

We can easily overestimate the importance of Praxagoras
and treat his teachings as the beginning of the "modern"
approach to medicine and physiology.[31] He does seem to
share the empirical approach of the unknown author of

[28] Of the 120 fragments and reports the bulk are found in the works of
Galen. This is especially true of the fragments relating to anatomy and
physiology.

[29] Steckerl, *Praxagoras*, pp. 33; 19, 36, 37; 12. Pneuma was acquired
from three sources: breathing, bubbles released by digestion, and respiration of the entire body surface.

[30] *Ibid.*, p. 11.

[31] See for instance Max Neuburger, *History of Medicine*, trans. Ernst
Playfair, vol. I (London, 1910), p. 167, who sees Praxagoras as "shaking"
the contemporary views upon respiration and establishing the foundation of later mechanical theories; also Steckerl, *Praxagoras*, p. 33, who
more cautiously suggests that Praxagoras may have had some vague idea
of oxidation and fermentation.

Ancient Medicine, but they were not alone in this outlook. Praxagoras, however, also shares in the tradition of Aristotle and Diocles and places the heart as the center of intellect, feeling, and life in general, even though the heart for Praxagoras was devoid of innate heat.[32] Suffice it to say that at least two early commentators were doubtful about the existence of an innate heat and the generalizations drawn from it.

Galen of Pergamon was the last and perhaps the greatest of the Greek men of medicine. His extensive writings formed the basis of medical thought well into the seventeenth century. The numerous editions of, and commentaries upon, his works were the texts from which generations of physicians learned the secrets of the body. Galen was also an important medical historian; writing in the second century A.D. he was able to look back and comment upon the teachings of his predecessors. His own doctrines are, in many ways, a synthesis of the findings of previous investigators.

Like his predecessors Galen raised the innate heat to a position of paramount influence. Not only does it preside over digestive and consequently growth functions, but heat is also responsible for evolving the psychic *pneuma* from the blood and thus indirectly controlling motion and sensation.

In regard to the nature of the body's heat, Galen has utilized the thoughts and criticisms of the older physicians. Food clearly emerges as a fuel for the body much in the manner that had been suggested by the Hippocratic authors and implied by the fragments of Praxagoras. Galen tells us that the fatty parts of the blood act as a kind of fuel for the heat of warm-blooded animals, whereas it is stored in

[32] Steckerl, *Praxagoras,* p. 36.

the colder ones. The very word nutriment implies increase in the heat of the animal, and conversely the lack of nourishment cools the body.[33]

An interesting analogy emerges in the discussion of food as a source of the internal warmth; Galen compares the production of heat from foods (particularly fatty foods) to the process of combustion as it occurs outside the body. The food "is used up by our heat as oil is by a flame." Or, "as oil nourishes an external fire, so the food of our nature which is like fire, is the fat of cooked foods."[34] The similarity between the vital heat and fire was more clearly drawn by Galen than by any previous authors. It seemed quite conceivable to Galen that an understanding of fire would lead to an understanding of the internal warmth. In discussing the relation of atmospheric air to flame and animal heat, Galen wrote:

Clearly we see these [flames], just as living things, swiftly extinguished when they are deprived of air. If a physician's cupping instrument or any narrow or concave vessel be put over the flames so as to cut off the access of air they are soon snuffed out. Now if we could discover why flames are in these cases extinguished, we should perhaps discover what advantage the heat in animals derives through respiration.[35]

[33] Galen (Kühn), I, 606, 660; XV, 265 as cited in C.-E. A. Winslow and R. R. Bellinger, "Hippocratic and Galenic Concepts of Metabolism," *Bull. Hist. Med.*, 17 (1945), 129. The citations of Galen are from the edition of Carl Gottlob Kühn, *Claudii Galeni Opera Omnia* (Leipzig, 1821–1833).

[34] *Ibid.*, p. 129–130, translated from Galen (Kühn), XI, 514; XVII, part 1, 745.

[35] Galen, from M. R. Cohen and I. E. Drabkin, *A Source Book in Greek Science* (Cambridge, Mass.: Harvard University Press, 1948), p. 256; (Kühn), IV, 487–488. Galen is comparing the innate heat to fire, not respiration to fire as is suggested by Wilson, "Erasistratus," p. 313. Galen does say that respiration seems to have a similar effect upon both a flame and the internal source of heat.

Unlike Aristotle, who was hesitant in drawing an analogy between fire and innate heat and sought at least once clearly to distinguish the two, Galen seems to find an almost complete identity between flame and the heat within the body. Not only are both phenomena nourished by fuel, they both utilize air in a process similar to respiration, and they also rid themselves of the waste products of combustion, soot, and ashes; "for ash and smoke and soot, and all superfluity of this sort of burned matter is wont, no less than water, to quench fire."[36] Respiration is able to conserve the internal warmth by "fanning of the source of the *innate heat* itself, and from its moderate cooling, and from an outflow of something like smoky mist."[37] This image of the heart containing a fire is one source of many misunderstandings in the later medical literature.

We will have created an erroneous impression, however, if Galen is viewed as a mechanist willing to have vital phenomena described wholly in terms of inanimate nature. The heat of the heart may be identified with fire, but none the less it was considered as inherent in the body. Galen had certainly not been in agreement with those physicians whom he noted as not believing in an innate heat and having devised, each in his own manner, an external origin of the body's warmth.[38]

The innate heat was considered necessary to the life of the animal; "digestion, nutrition, and the generation of the various humours, as well as the qualities of the surplus

[36] Winslow and Bellinger, "Concepts of Metabolism," p. 135; Galen (Kühn), IV, 492, also translated in Wilson, "Erasistratus," p. 313.

[37] Galen, in Wilson, "Erasistratus," Kühn, IV, 492. See also Donald Fleming, "Galen on the Motions of the Blood in the Heart and Lungs," *Isis,* 46 (1955), 18.

[38] Steckerl, *Praxagoras,* p. 57, fragment Galen (Kühn), VII, 614.

substances, result from the *innate heat*."[39] Galen consciously joined the tradition of Hippocrates and Aristotle
when noting that these views were shared by his illustrious
predecessors. The heart was considered the source of the
vital heat[40] and from that organ the pulse distributed
warmth to various parts of the body.[41]

Respiration, according to Galenic physiology, has conservation of the innate heat of the heart as its primary
function. A similar function for respiration had been proposed by Aristotle. The Peripatetic view suggested that
conservation was achieved because breathing served to
refrigerate and moderate the internal heat. Galen, on the
other hand, intimates that the heart requires the substance
of the air as well as the refreshing cooling; and he also
adds that the breath in passing out of the body carries with
it the burnt and smoky particles alluded to above. The
substance of the air, however, seems to be involved in the
formation of the vital spirit and not in nourishing the heat,
for Galen contrasts his own view with that of Hippocrates,
who believed that respiration was useful for "nutrition" as
well as "refreshing cooling."[42]

Galen suggests that his predecessors derived their theories
of the function of respiration and its relation to innate heat
through analogy with "the facts which appear in connec-

[39] Galen, *On the Natural Faculties*, trans. A. J. Brock (Loeb Classical
Library; Cambridge, Mass.: Harvard University Press; London: Heinemann, 1947), p. 141.

[40] Galen, *On Anatomical Procedures*, trans. C. Singer (London: Oxford University Press, 1956), p. 184.

[41] Wilson, "Erasistratus," p. 305; Galen (Kühn), V, 161.

[42] *Ibid.*, pp. 306, 311, Galen (Kühn), III, 412; IV, 471. Wilson, p. 313,
suggests that "air served to cool this *innate heat* as well as nourish it."
I am not sure how he arrives at the notion of nourishing since all his
arguments would suggest that the nutriment of the heat came only from
food.

tion with flames."[43] That this is the method utilized by Galen himself is made amply clear by his many references to the requirements of fire.

Although respiration is the main agency through which the vital heat is conserved, Galen also turned to the brain as a moderating influence upon the heat of the heart. His reasons, however, are different than those of Aristotle, for whom the brain was naturally cold. Galen concludes that the brain which is by nature susceptible to injury from extremes of heat or of cold has been provided with the power to control the respiration and thus moderate the heat generated in the heart.[44]

The innate heat as Galen used it is conceptually similar to the theory as it appears in the works of other Greek physicians and philosophers.[45] Heat is considered inherent in the body, its origins unclear, but closely tied to life itself. The respiratory process served in one manner or another

[43] *Ibid.*, p. 312, Galen (Kühn), IV, 487.

[44] Winslow and Bellinger, "Concepts of Metabolism," p. 133, Galen (Kühn), II, 884.

[45] I have not attempted to root out all the Greek theories of animal heat but have concerned myself with several major trends. It would prove interesting to explore the views of the atomists; see, for example, Cyril Bailey, *The Greek Atomists and Epicurus* (Oxford, 1928), pp. 156ff, 384ff, who notes that for Leucippus and Democritus "the soul or vital principle was corporeal, and that it was of the nature of fire," p. 156. For Stoic concepts of vital heat see G. Verbeke, *L'Evolution de la Doctrine de Pneuma du Stoicisme à Saint Augustin* (Paris, Louvain: Desclée de Brouwer, 1945), and also S. Sambursky, *Physics of the Stoics* (London: Routledge and Kegan Paul, 1959). For the Stoics, *pneuma* and the vital heat are identified with fire, life being dependent upon them. For additional, often uncritical, views of the early theories of innate heat, see T. C. Albutt, *Greek Medicine in Rome* (London, 1921), pp. 239–262, and "The Innate Heat," *Contributions to Medical and Biological Research, Dedicated to Sir William Osler* (New York, 1919), I, 219–225. An important source for the relations of Diocles and the Peripatetics is Werner Jaeger, *Diokles von Karystos, die Griechische Medizin und die Schule des Aristoteles* (Berlin: de Gruyter, 1938).

to preserve, cool, or moderate the heat at its source, the heart. In varying degrees the ancients recognized a need to nourish or provide fuel for the internal fire. The similarity to external flame was recognized and exploited, but care was generally taken to differentiate the vital heat from ordinary combustion. Perhaps more important than any other consideration was the conception of the innate heat as a major motive power responsible for a wide range of vital phenomena, a motive power considered to be generally outside the range of physical causation.

As I have indicated, the doctrine of innate heat became so important a part of physiological thought that it was not until the seventeenth century that it was successfully challenged and replaced. The lasting power of the theory was certainly strengthened by the close association it had with the traditions of Hippocrates, Aristotle, and Galen. A few samples of the manner in which innate heat was treated in the long span of years from antiquity to the seventeenth century lend weight to the contention that it underwent very little basic change during this period.

Nemesius, Bishop of Emesa, a late-fourth-century Christian philosopher, was the author of an important book *On the Nature of Man*.[46] The work, which was widely appreciated during the early Middle Ages, was the source of much physiological knowledge. In describing the function of respiration Nemesius gives an account of the innate heat which is strikingly similar to Galen's. The heart was considered the source of the natural heat. Respiration, both inhalation and exhalation, serves to maintain the internal warmth; "While inhalation first cools and then gently fans

---

[46] Nemesius, *Of the Nature of Man* in *Cyril of Jerusalem and Nemesius of Emesa*, ed. William Telfer (Library of Christian Classics, IV; Philadelphia: Westminster Press, 1955). See also George Sarton, *Introduction to the History of Science* (Baltimore, 1927), I, 373–374.

up the vital heat, exhalation pours forth the smoky fumes of the heart." The lungs, by enveloping the heart in their midst, are able to provide the needed cooling. "The natural heat of life" is distributed to all parts of the body from the left ventricle of the heart through the pulsating arteries. "So long as the heart is hot, just so long is the whole living body heated."[47]

No new elements have been added by Nemesius to the doctrine of the vital heat, but this Christian author has given testimony to his familiarity with the prevailing medical theories.

In the early eleventh century one of the most important medical texts of all time, the *Canon of Medicine*, provided a codification of ancient and Muslim medical knowledge. Avicenna, the author of this immense encyclopedia, had been strongly influenced by Galen; consequently, as might be expected, the subject of innate heat is quite prominent in the pages of the *Canon*. The internal heat is equated with vitality, and differentiated from foreign, or external, heat which is harmful to bodily health. "If the *innate heat is strong*, the natural faculties are able to work through it, upon the humours, and so effect digestion and maturation, and so maintain them within the confines of the healthy state."[48] Weakness of the internal warmth, on the other hand, allows external agents to dominate and direct the system, thus leading to "putrefaction."

Avicenna continues an important, if often confusing, tradition when he links exercise to an increase in innate heat.[49] Muscular activity produces an obvious or sensible

---

[47] Nemesius, *Nature of Man*, pp. 376, 377, 366.
[48] O. Cameron Gruner, *A Treatise on the Canon of Medicine of Avicenna, Incorporating a Translation of the First Book* (London, 1930), pp. 129, 270.
[49] *Ibid.*, p. 210.

increase in bodily warmth. The confusion often arose as later interpreters tried to use this as proof that innate heat was generated by a mechanical process. However, in *Regimen* II, Hippocrates links exercise not only to an increase in heat but also to increased concoction and digestion.[50] Galen had also noticed that "excited motion" increases the body heat, while immobility causes it to languish.[51]

Jean Fernel, in the early sixteenth century, was anxious to clarify the distinction between fire and innate heat, a point indecisively broached by Aristotle. "Innate heat," said Fernel, "is not of the same nature as fire. It comes from a different source from fire." He argued that life lives by the innate heat, not by fire or the elemental heat. The vital heat enters the animal at the very moment that the embryo becomes an independent individual, that is, on about the fortieth prenatal day.[52]

The starry sphere is the source of the innate heat; thus it is not derived from elemental heat or fire, but rather from the quintessence, or aether, and therefore also resembles the heat of the sun.[53] This is precisely the distinction that Aristotle was trying to make in *De generatione animalium;*[54] the "hot substance" was not fire, but rather a substance analogous to the element which belongs to the stars.[55] Death, Fernel points out, is the extinction of the innate heat, the vital heat. "Death demonstrates that this heat is no result of a mere mixture of the elements. Death

[50] Winslow and Bellinger, "Concepts of Metabolism," p. 130.
[51] *Ibid.*, p. 131, Galen (Kühn), VII, 772.
[52] As translated in Sir Charles Sherrington, *The Endeavour of Jean Fernel* (Cambridge, Eng.: University Press, 1946), pp. 38–40.
[53] *Ibid.*, p. 39.
[54] Aristotle, *De generatione animalium*, 736 b 35.
[55] See Solmsen, "The Vital Heat," p. 119.

occurs and still the body retains its elements and the shape of all its parts. We recognize our friend although his life is not there. The innate heat has fled him."[56] In a further distinction Fernel suggests that it is by virtue of the innate heat that a snake lives, even though its temperament is cold. The same argument is carried over to plants. Innate heat then is inseparable from life.

The distinction between the two varieties of heat is not accomplished without confusion, however, even by Fernel. The implication is that innate heat might exist without sensible heat. This image is not maintained, for when Fernel turns to digestion, which he claims is assisted by the innate heat, he compares this heat to a fire put under a cauldron. He says also that a finger inserted into the left ventricle of the heart (the source in the body of innate heat) finds this organ the hottest place in the whole body. But he also claims that the reason shellfish, which have neither warmth nor blood, can perform digestion is that the heat employed was innate heat.[57]

As we can readily see from this discussion of Fernel, had the criteria of sensibility of heat been employed in all of the foregoing discussion of theories of innate heat the confusion would have been compounded. In one case the vital heat is compared to fire, in another it is believed to exist where there is no sensation of warmth. Part of the difficulty certainly stems from the special properties and characteristics attributed to the innate heat. But another, and equally fundamental, side of the problem was the inability, in early discussions of heat, to distinguish between intensity and quantity of heat. This is not to say that this fundamental

[56] Sherrington, *Jean Fernel*, p. 39.
[57] *Ibid.*, p. 69.

distinction had not been made, but rather that in the discussions of heat in the animal body the differentiation had not been utilized.[58]

[58] Marshall Clagett, *Giovanni Marliani and Late Medieval Physics* (New York: Columbia University Press, 1941), provides an excellent discussion of the attempted distinction between intensity and quantity of heat. Galen had attempted a classification of medicinal simples in a system of four orders or degrees of hot and cold. He relied upon the senses to distinguish degrees of "sharpness" or "pepperiness" and correlated these with heat. (Galen, *De simplicium medicamentorum temperamentis ac facultatibus,* III, 13, Kühn, XI, 570ff.) Clagett deals at length with medieval discussions of the quantity of heat and includes a chapter on physiological heat. Giovanni Marliani had accepted a distinction between "intensively hot" and "naturally hot" (p. 92). A body is absolutely (intensively) hotter than another when it has a more intense degree of heat. A body is hotter according to nature when it has a greater quantity of blood, natural heat, and *spiritus* (p. 87). We will turn, at greater length, to the question of heat and temperature in Chapter V.

·III·

Prototheories

Within this city is the palace fram'd,
Where life, and life's companion, heat, abideth;
And there attendants, passions all untam'd:
(Oft very hell, in this straight room resideth)
And did not neighbouring hills, cold airs inspiring,
Allay their rage and mutinous conspiring,
Heat, all (itself likewise) wou'd burn with
    quenchless firing.
                    —Phineas Fletcher, *The Purple Island*[1]

The seventeenth century was a period of transition and contrasts in science. This judgment, which has become a commonplace in discussions of physics, astronomy, and cosmology, is not quite as clearly asserted in relation to chemistry or biology. After all, the revolution which we recognize as having occurred in the physical sciences brought into being a mechanics of the macrocosm and microcosm which is familiar even to the modern scientist. This is not the case for chemistry or biology. The concepts

[1] Phineas Fletcher, *The Purple Island or the Isle of Man. An Allegorical Poem* (London, 1783), new edition, Canto IV, 25, originally published 1633.

and theories developed in these two fields[2] during the seventeenth century have long since been replaced. Nevertheless, the changes which took place in biology, for instance, seem no less revolutionary than those in the exact sciences. This can be maintained, I believe, because the major innovation of the "scientific revolution" was one of scientific attitude rather than of scientific invention. This change of attitude is well reflected in the biological sciences and can be particularly well illustrated by a look at the transition which occurred in the theory of animal heat.

The theory of an innate body heat, as shown in the previous chapter, was developed by the early Greek physicians and philosophers into a compact doctrine through which a multitude of physiological functions could be explained. The power and authority of the theory of innate heat was such that it was utilized, in substantially unchanged form, for almost two millenia. There were several ancient critics and one or two in later time;[3] but these aside, the theory, and the Galenic medical teachings in which it

---

[2] Marie Boas, in *Robert Boyle and Seventeenth-Century Chemistry* (Cambridge, Eng.: University Press, 1958), has carefully studied the new developments which took place in chemistry during the period of the "scientific revolution."

[3] The notebooks of Leonardo da Vinci indicate his objection to considering the body heat as inherent: "The heat is produced by the movement of the heart, and this manifests itself because in proportion as the heart moves more swiftly the heat increases more." Edward MacCurdy, ed., *The Notebooks of Leonardo da Vinci* (New York: Reynal and Hitchcock, 1939), p. 128. But this is one of the few deviations that Leonardo makes from Galenic doctrine. He accepted the view of the heart as the source of heat and believed with Galen that heat is responsible for motion and is necessary for life. See K. D. Keele, *Leonardo da Vinci on Movement of the Heart and Blood* (London: Harvey and Blythe, 1952), pp. 125–126, 93–94, 19. Leonardo's writings, however, were not published until the nineteenth century.

played an important part, were accepted by the major medical writers into the seventeenth century.[4]

William Harvey, the celebrated physiologist, stood at the threshold of modern science, and a good case can be made for considering Harvey as an early apostle of experimentation.[5] "I have not wanted," he claimed, "to demonstrate from causes and probable beginnings but, through sense and experience, in anatomical fashion I have wished, as if by higher authority, to bring to confirmation."[6] But it was the same Harvey who appealed to the authority of the ancient masters for his belief in the innate heat. In a manner reminiscent of Aristotle he claimed that the heart "deserves to be styled the starting point of life and the sun of our microcosm just as much as the sun deserves to be styled the heart of the world . . . The heart is the tutelary deity of the body, the basis of life, the source of all things, carrying out its function of nourishing, warming, and activating the body as a whole."[7]

Harvey's ideas about the place and function of heat in the

[4] Andreas Vesalius, whose anatomical studies were one of the major scientific advances of the mid-sixteenth century, remained wholly within the Galenic tradition in physiology. The heart is the center of an innate heat; respiration serves to ventilate and also remove the smoky wastes of the vital heat; the blood carrying innate heat tempers and restores the parts of the body. See Andreas Vesalii, *De humani corporis fabrica* (Basel, 1543), Book VI, Chap. 1.

[5] For a provocative discussion of this point see J. A. Passmore, "William Harvey and the Philosophy of Science," *Australasian Journal of Philosophy*, 36 (1958), 85–94.

[6] William Harvey, "A Second Essay to Jean Riolan, in which many objections to the circuit of the blood are refuted," in *The Circulation of the Blood, Two Anatomical Essays by William Harvey together with nine letters written by him,* trans. and ed. Kenneth J. Franklin (Oxford: Blackwell, 1958), pp. 58–59.

[7] William Harvey, *Movement of the Heart and Blood in Animals, An Anatomical Essay,* trans. Kenneth J. Franklin (Oxford: Blackwell, 1957), p. 59.

body underwent a change during his lifetime. In the lecture notes of 1618 (among the earliest extant writings) Harvey is very much in the ancient tradition. This is also true when he is not specifically dealing with the question of heat, but rather treating it incidentally to some other discussion. The later writings, however, exhibit a more cautious use of the notion of innate heat and show a generally more sophisticated approach in dealing with the living organism.

Harvey's early notebooks stress the supreme importance of heat for life of the animal, and of the necessity of cooling and ventilation to the continuance of the internal heat.[8] The heat, he believed, could perish either by wasting or smothering.[9] Heat is necessary in the concoction or digestion of food and hence to the nourishment of life. The movement of animals was also shown to be effected by the internal warmth, in that "the power, the capacity for tension, etc., of a muscle derive from spirit and from heat."[10] "Warmer things," he believed, "are more agile and colder things more lazy;" a rule which he believes to hold true for all living creatures and for every individual of every species.[11]

As he outlines his doctrine of the circulation of the blood

[8] Cited in John G. Curtis, *Harvey's Views on The Use of the Circulation of the Blood* (New York, 1915), p. 19, from *Prelectiones Anatomiae Universalis*, edited with autotype reproduction of the original by a Committee of the Royal College of Physicians (London, 1886); see *prelectione* 86. See also the recently prepared volume, William Harvey, *Lectures on the Whole of Anatomy, An Annotated Translation of "Prelectiones Anatomiae Universalis,"* trans. and ed. C. D. O'Malley, F. N. L. Poynter, and K. F. Russell (Berkeley: University of California Press, 1961).

[9] Compare this with reports of Aristotle and Galen in Chapter II.

[10] William Harvey, *De Motu Locali Animalium* (1627), ed. and trans. with intro. by Gweneth Whitteridge (Cambridge, Eng.: University Press, 1959), p. 129. In this early work Harvey is still associating heat with spirit. Later, under prodding from Descartes, he attempts a clear distinction.

[11] *Ibid.*, p. 103. These notes, which remained unpublished in Harvey's lifetime, give ample evidence of Harvey's debt to ancient authority, Aristotle in particular.

Harvey demonstrates the importance of the innate heat to his physiological theory. If there is a *reason* for the circulation—and Harvey is loath to speculate about the reason for bodily function[12]—it is that through the movement of the blood the peripheral parts are warmed and nourished and in essence have life brought to them.[13] Harvey was quite aware, as Galen had been before him, that if the arteries were tied off the part they normally served would become cold and lifeless.[14] Thus as the blood moves through the extremities it is cooled and thickened and has to return to the heart to be warmed and revivified in this organ which Harvey considered as the site of the *innate fire* and the *beginning of life*.[15]

A much better understanding of what Harvey meant by *circulation* of the blood is gained by examining the role he attributed to heat. Harvey had in mind a *cyclic* process of the heating, evaporation, and subsequent cooling down and condensation of a fluid.[16] Heat is the operative factor, causing the evaporation which is followed by cooling.

It may very well happen thus in the body with the movement of the blood. All parts may be nourished, warmed, and activated

[12] "One ought to admit what should be investigated rather than the reason for such further study." Harvey, "Second Essay to Riolan," p. 45.

[13] Harvey, *Movement of the Heart and Blood*, pp. 59, 88.

[14] *Ibid.*, p. 11.

[15] *Ibid.*, pp. 88, 89: "Unless, too, the heart were indeed such a centre as would retain life and warmth with the extremities frozen (see Aristotle, *De respiratione*, Cap. 2), and would transmit through the arteries fresh, warm, spirit-imbued blood to drive out the chilled and effete matter, and to allow the parts to become warm again and revive their all but extinguished vital fire?"

[16] "We have as much right to call this movement of the blood circular as Aristotle had to say that the air and rain emulate the circular movement of the heavenly bodies. The moist earth, he wrote, is warmed by the sun and gives off vapours which condense as they are carried up aloft and in their condensed form fall again as rain and remoisten the earth." Harvey, *Movement of the Heart and Blood*, p. 58.

by the hotter, perfect, vaporous, spirituous and, so to speak, nutritious blood. On the other hand, in parts the blood may be cooled, coagulated, and be figuratively worn out. From such parts it returns to its starting-point, namely, the heart, as if to its source or to the centre of the body's economy, to be restored to its erstwhile state of perfection.[17]

A similar cycle of warming and cooling occurs in the "lesser circulation." The blood in passing from the right ventricle of the heart is cooled and saved from "bubbling to excess" by its contact with the inspired air in the lungs before being returned to the left ventricle.[18] Harvey finds in these cyclic processes the means by which the two extremes of hot and cold are balanced, thus permitting the maintenance of the normal temperature of the animal.[19]

As Harvey turns his discussion, in his later writings, to the nature of the innate heat there is a noticeable change in attitude. Heat is still present in the body; it is one of the "brute facts" of higher organisms. But a new note of caution has been added, for even though he is still willing to "assert the native heat, or innate warmth, to be the common instrument of all operations, and also the primary efficient cause of the pulse," he quickly adds, "this I do not as yet constantly assert, but merely propose as a thesis."[20]

Although in earlier writings Harvey was careless in his characterization of the internal heat, often comparing it to ordinary flame or fire, he now wished to distinguish the

[17] *Ibid.*, p. 59. See also Walter Pagel, "William Harvey and the Purpose of Circulation," *Isis*, 42 (1951), 22–24. Also Étienne Gilson, *Études sur le Rôle de la Pensée Médiévale dans la Formation du Système Cartésien* (Paris: Vrin, 1951), p. 72.

[18] Harvey, *Movement of the Heart and Blood*, p. 50.

[19] Harvey, "The First Anatomical Essay to Jean Riolan on the Circulation of the Blood," *Circulation of the Blood*, p. 20.

[20] Harvey, "Second Essay to Riolan," p. 63. This work, written partially as a reply to Descartes's anatomical observations, is also the most consciously philosophical of Harvey's treatises.

two carefully. Citing Aristotle as his authority, Harvey maintained that the innate heat is not fire nor does it derive from fire; "that the heat of the blood of animals during their lifetime, therefore, is neither fire, nor derived from fire, is manifest, and indeed is clearly demonstrated by our observations."[21] Harvey proceeds to link the internal heat closely with blood; it has no existence apart from blood, and blood without heat "is no longer blood, but cruor or gore." There are no other spirits of an "aerial or ethereal nature . . . more excellent and divine than the innate heat," as had been claimed by Scaliger and Fernel in the sixteenth century. The blood itself is the only *calidum innatum* and thus there is "no occasion for searching after spirits foreign to, or distinct from the blood." There is no reason whatever, Harvey concludes, "to evoke heat from another source; to bring gods upon the scene, and to encumber philosophy with any fanciful conceits; what we are wont to derive from the stars is in truth produced at home."[22]

But lest we leave Harvey sounding much more modern than he really was we must follow him somewhat further. Heat and blood, which are neither fire nor the product of some spirit, do share *something* which surpasses the power of the terrestrial elements; it is the same *something* which inheres in the semen and makes it prolific. This *something* is identical with the essence of the stars. But with all this said, Harvey wants to make us sure that all the things which have been attributed to the spirits or the innate heat belong to the blood alone. "We are too much in the habit, neglecting things, of worshipping specious names. The word blood, signifying a substance, which we have before our eyes, and can touch, has nothing of grandiloquence about

21 William Harvey, *Anatomical Exercises on the Generation of Animals*, in *The Works of William Harvey*, trans. Robert Willis (London, 1847), pp. 505–506.
22 *Ibid.*, pp. 502–503.

it; but before such titles as spirits, and *calidum innatum*, or innate heat, we stand agape."[23]

Having identified the innate heat with the blood, making heat as characteristic of blood as contraction is of muscle, Harvey proceeded to reduce the role of the heart. He no longer considered it as the fashioner of the blood, or as the source of the innate heat of the blood. The heart is not "like a sort of burning coal or brazier or hot kettle, the source of heat and blood, but rather the blood, as being the warmest part of all in the body, gives to the heart (as to all the other parts) the heat which it has received." Furthermore, Harvey argues, the reason that the heart has been considered as the "warehouse, source, and permanent fireplace . . . like a hot kettle," is solely that it contained more blood and "not by reason of its fleshiness." For the very same reason, their large supply of blood, he added, the liver, the spleen, and the lungs have been called hot parts.[24]

Harvey's changing notions about the innate heat are directly related to his development as an "experimental philosopher." Although the heat of the blood still had a touch of the divine, Harvey had in essence reduced it to a property (indeed an important one) of the blood. He did not offer any real explanation of the heat, nor did he seek any physical cause for it; but then, he had become justly cautious of explaining and seeking causes. Harvey was sure of the fact that heat was in the body, and that it was closely tied to the maintenance of life; he was willing to describe the role of the heat of the body, but closed his discussion of the innate heat with the caution that what he said was proposed "as a thesis."

One of the earliest defenders of Harvey's theory of

[23] *Ibid.*, pp. 506f, 510–511.
[24] Harvey, "Second Essay to Riolan," p. 63.

circulation of the blood was René Descartes. In a letter to Mersenne in 1632[25] Descartes notes having read the *De motu cordis* and having already discussed some of the questions involved "en mon Monde."[26] His first published pronouncement in favor of the Harveian theory appears, however, in part five of the *Discourse on Method*.[27] Upon reading this discussion it is immediately apparent that Descartes's support for the theory of the circulation does not extend to Harvey's explanation of the motion of the heart.[28] Instead of relying, as does Harvey, upon the contractile nature of the heart muscle, Descartes introduces a theory which depends upon an intense heat in the heart. This proposal, which has often been viewed as Descartes's mistaken interpretation of the physiology of the *De motu cordis*, is actually an alternate conception growing out of his attempt to construct a physiology consistent with his general view of the mechanics of nature.

Descartes was particularly proud of his explanation of the heart's motion; he believed that he had been able to provide a purely geometric and mechanical explanation of an important "vital" phenomenon.[29] After all, the beating

[25] Descartes to Mersenne, Nov. or Dec. 1632, *Oeuvres de Descartes*, ed. Charles Adam and Paul Tannery (Paris, 1897–1910), I, 260–263.

[26] *Ibid.*, p. 263. This is in reference to the section of *Le Monde* called *Traité de l'Homme*. Although written in 1632, *L'Homme* was not published until 1664. It has been republished *in toto* in the *Oeuvres*, XI, 119–215.

[27] Descartes in referring to the circulation speaks of "what has already been written by an English physician, to whom the credit of having broken the ice in this matter must be ascribed." *Discourse on the Method of Rightly Conducting the Reason and Seeking for Truth in the Sciences*, in *The Philosophical Works of Descartes*, trans. E. S. Haldane and G. R. T. Ross (Cambridge, Eng., 1931), I, 112, *Oeuvres*, VI, 50.

[28] *Discourse*, pp. 114f, *Oeuvres*, VI, 54f. In the letter to Mersenne (*supra*) Descartes had already indicated some difference of opinion with Harvey.

[29] Gilson, *Système Cartésien*, p. 51.

of the heart is "the first and most general movement that is observed in animals."[30]

The process Descartes described could easily be based upon the analogy from Aristotle that we found in Harvey's *De motu cordis*.[31] Descartes believed that blood upon entering the ventricles of the heart, drop by drop, was caused "to expand and dilate, as liquids usually do when they are allowed to fall drop by drop into some very hot vessel."[32] The blood thus expanded and rarefied is driven into both the pulmonary artery and the aorta. Respiration serves to provide sufficient fresh, cool air to cause the vaporized blood, which has entered the lungs from the right ventricle, "to thicken, and become anew converted into blood before falling into the left cavity," where it serves as fuel for the fire in the heart. The blood driven out through the aorta into the arterial system provides warmth to the body, and is particularly needed in the stomach to help digestion.[33]

In this explanation of the heart's motion Descartes was sure he had found an excellent illustration for "those who do not know the force of mathematical demonstration and are unaccustomed to distinguish true reasons from merely probable reasons." He was confident that the movement "just explained follows as necessarily from the very disposition of the organs, as can be seen by looking at the heart, and from the heat which can be felt with the fingers . . . as does that of a clock from the power,

[30] Descartes, *Discourse*, p. 110, *Oeuvres*, VI, 46.

[31] See Harvey's use of the analogy above, note 16. Plempius in a letter to Descartes called his attention to Aristotle's explanation of the movement of the heart by heat. Plempius à Descartes, Jan. 1638, *Oeuvres*, I, 497.

[32] Descartes, *Discourse*, p. 111, *Oeuvres*, VI, 48–49.

[33] *Discourse*, p. 114, *Oeuvres*, VI, 53.

the situation, and the form, of its counterpoise and of its wheels."[34]

According to the Cartesian view, the heart is little more than a container within which the blood is heated. And the source of heat? What better cause could there be "of very great and sudden changes in the body except boiling and fermentation?"[35] This process is elsewhere referred to as "un de ces feux sans lumière," in no way different from that which causes hay to heat up and makes new wine grow frothy.[36]

It is this very notion that Harvey was criticizing when he insisted that the heart's motion was due to the contractile nature of muscle. The very type of movement found in the heart would rule out Descartes's proposal; there is nothing, wrote the English physician, "in fermentation or bubbling up [which] rises and falls as if in the winking of an eye."[37] But Descartes was far more willing to depend upon such well-known and generalized phenomena as fermentation and boiling ("fire without flame") than to rely, as he felt Harvey did, upon the contractile "faculty" of muscle. "If we suppose that the heart moves in the manner in which Harvey described it," complains Descartes, "we shall have to imagine some faculty which causes this movement, the nature of which is much more difficult to conceive than everything he claims to explain by it; it also will be necessary to suppose other faculties which change the quality of the blood while in the heart."[38]

---

[34] *Discourse*, p. 112, *Oeuvres*, VI, 50. Note that Descartes claims to have felt the heat of the heart.

[35] Descartes à Plempius, 15 Feb. 1638, *Oeuvres*, I, 531. See also Passmore, "William Harvey," p. 90.

[36] Descartes, *Discourse*, p. 109, *Oeuvres*, VI, 46; *Traité de l'Homme*, *Oeuvres*, XI, 123.

[37] Harvey, "Second Essay to Riolan," p. 66.

[38] Descartes, *La Description du Corps Humain*, *Oeuvres*, XI, 243–244. Descartes was critical of Harvey for not being able to ascribe a cause to

But in order to cast out the "faculties," which he considered a holdover from scholastic philosophy, Descartes was willing to have posited in the heart a heat which would provide motion for the heart. What he seemed not to realize was that the very same type of fire which he had God enkindle in the heart[39] was not only found in ancient philosophy but also in his immediate predecessors, the Scholastics. The idea of fire was central to the whole of Descartes's physical philosophy and probably for this reason he was willing to risk identification with Scholastic physiology.[40] The element fire accounts for the sun and the stars, for combustion and the flame which accompanies combustion, and now he has proclaimed its adequacy to account for animal heat, and through this for life as it is found in animals and man.[41]

With the helpful intervention of fire or heat Descartes believed he was successful in treating the living body as a machine; heat is assigned as the cause of all the movements of this machine and the heart is considered as the motor or source of the heat:[42] "so long as we live there is a continual fire in our heart, similar to the fire found in the blood of the vessels, the fire which is the basis of all the movements of our parts."[43] This fire in the heart is outside

the change of color of the blood in the lungs. Descartes attributed the change to the rarefaction and subsequent condensation that the blood undergoes in the heart and lungs.

[39] Descartes, *Discourse*, p. 109, *Oeuvres*, VI, 46.

[40] Gilson, *Système Cartésien*, pp. 83, 64–73.

[41] Norman Kemp Smith, *New Studies in the Philosophy of Descartes* (London: Macmillan, 1952), p. 129. Smith cites a letter from Descartes to Henry More, Feb. 6, 1649, *Oeuvres*, V, p. 278: "Life I deny to no animal, except in so far as I lay it down that life consists simply in the warmth of the heart." (p. 126)

[42] See Aug. Georges Berthier, *Le Mécanisme Cartésien et la Physiologie au XVIIe Siècle* (Brussels, n. d.), p. 16.

[43] Descartes, *Les Passions de l'Ame*, *Oeuvres*, XI, 333. This is the second place where Descartes talks of the blood maintaining or nourishing the fire in the heart; see note 33 above.

the control of the soul or spirits.[44] In fact, the warmth of the heart is responsible for the generation of the animal spirits "which resemble a very subtle wind, or rather a flame which is very pure and very vivid," and which continually rises from the heart to the brain and from there through the nerves to the muscles thereby providing the power of motion to the body's parts.[45]

Descartes may have been satisfied with his description of the human body as a machine, a machine powered by the intense heat maintained in the heart. But the heat utilized is conceptually similar to the innate heat proposed by Aristotle. It is similar in the breadth of the functions for which it is considered responsible, and in that no cause is offered to explain its existence. Descartes used the heat to run a machine rather than to integrate the activities of an organism, but the result is the same; a heat is believed to exist in the heart, the specific presence of which cannot be experientially verified, the specific cause of which is not sought.

Whereas Descartes was anxious to treat the whole organism as a machine, the iatrochemists of the seventeenth century tried to apply chemical theory to various individual physiological functions. This approach is certainly, in a limited way, mechanistic, for it involves an attempt to attribute to living processes the concepts, laws, and mechanisms utilized in explaining the nonliving world.[46] This new outlook is particularly important in a study of the changing ideas about the warmth of the animal body. It could no longer be assumed, as it had been, that heat inhered in the living body, or that heat was a generalized power

---

[44] *Ibid.*, p. 329.
[45] Descartes, *Discourse*, p. 115, *Oeuvres*, VI, 54.
[46] Boas, *Robert Boyle*, p. 58, has suggested that this is often hard to realize since, for an important figure like J. B. van Helmont, "his nonliving world often sounds curiously animate."

governing a whole array of physiological activities. Each function was examined and a chemical reaction found to explain it.

Harvey and Descartes might be considered as joint forebears of this new approach; Descartes had insisted on generalized mechanical explanations of vital phenomena, and Harvey had turned to the specific process and demanded that it be carefully examined and understood for itself. An important addition to the new methodological contributions of Harvey and Descartes was the introduction of the belief that physiology could be explained in terms of chemistry. A pioneer in these attempts to develop a chemistry of the living system was J. B. van Helmont.[47]

Van Helmont rejected the idea that innate heat is the chief vital factor, or is even essential to life. After all, he argued, frogs and fishes are cool but life in them is as real as in the lion. Death of the warm-blooded animal comes not because warmth cools off, but rather warmth disappears because an animal dies. The heart of a cold-blooded frog beats in a manner similar to those of other animals, making it quite clear that there is no innate heat in the heart responsible for its expansion and relaxation. Heat, van Helmont might have summed up, is a companion of life, a symbol of the mode of operation in warm-blooded animals; it is not the determining factor.[48] His argument is all the more interesting because it demonstrates the immediate power of comparative studies in biology. Although statements of a comparative nature are found scattered through the writings of earlier naturalists, the seventeenth century

[47] Boas, *ibid.*, p. 55, notes that Helmont is a difficult scientist to assess because he is often mystical and obscure in his writings.

[48] Walter Pagel, *The Religious and Philosophical Aspects of van Helmont's Science and Medicine, Supplements to Bull. Hist. Med.*, No. 2 (Baltimore, 1944), p. 4.

marks the period during which comparative studies became an essential part of biology. By turning directly to nature the scientist found new elements for the construction of his explanations.

Van Helmont criticized the ancients for attributing to heat the power of governing growth and generation. Warmth, he believed, might hasten the hatching of an egg, but it could never be responsible for the production of the chick. Heat can generate nothing but more heat.[49]

In his study of digestion van Helmont made what are probably his most important contributions to physiological chemistry. Again, he found the ancients (and with them the Schoolmen) in error by their supposing that heat was the agent of digestion.[50] The internal warmth may well support digestion, but for van Helmont acid is definitely the operative agent.[51] Fish, he pointed out, which are devoid of animal heat, have a very efficient digestive system. People were probably misled into considering heat the important factor through their observation of the softening of meat

[49] *Ibid.*, p. 5.

[50] Van Helmont's criticism of the role of heat in digestion was in part anticipated by earlier Renaissance figures. Johannes Reuchlin (1455–1522) expressly rejected heat as the decisive agent in the animal economy. Digestion, he believed, could not be carried out by internal heat, for, if it could, heat or fire external to the body should be able to digest food better and more rapidly than the process in the stomach. He believed that a specific (occult) virtue must effect the gastric digestion and that heat, which is a general factor, plays no more than an auxiliary role, acting as a medium to support a function essentially different from it. Walter Pagel, *Paracelsus, An Introduction to Philosophical Medicine in the Era of the Renaissance* (Basel: Karger, 1958), p. 293. Paracelsus (of whom van Helmont might be considered a disciple), adopted a similar view. A specific process was involved in gastric digestion. There exists in the stomach " 'a mighty heat which so efficiently seethes and cooks—not unlike the fire outside.' " It is this heat which distributes itself to the organs, thus providing the heat of the body. *Ibid.*, p. 159.

[51] Walter Pagel, "Van Helmont's Ideas on Gastric Digestion and the Gastric Acid," *Bull. Hist. Med.*, 30 (1956), 527.

upon cooking. But this, van Helmont complained, could hardly be considered as true digestion, since the structure of the meat remained unchanged. True digestion involves the transmutation of food into the parts of the body and this cannot be achieved by even the highest degree of heat.[52]

Heat, van Helmont pointed out, is a general factor and therefore should evidence the same action in whatever case applied. How, therefore, could heat possibly serve as the cause of such diversified results as those occurring in digestion? In adopting this point of view van Helmont expressed ideas at variance with most of the men who had previously studied digestion. They had been quite pleased at finding one factor that could produce widely variant results; it was not so for van Helmont. He believed that the constituent parts of each animal species were distinct and that their formation must be explained in terms of a multiplicity of specific causes or properties each closely tied to a single species or even an individual. Heat was a nonspecific force, with all species and individuals partaking of it in a uniform manner.[53]

Having ruled out heat as the active agent of digestion, van Helmont turned to what he considered a purely chemical process, fermentation. Digestion, he thought, took place in the stomach, the liver, and other organs by means of an acid fermentation, not unlike that which turns grapes into wine.[54] Fermentation was believed by van Helmont to be the kind of chemical reaction capable of producing heat or effervescence.[55]

[52] Jean Baptista van Helmont, *Oriatrika or, Physick Refined*, trans. J. C. (London, 1662), pp. 198–203. See also Pagel, "Van Helmont, Gastric Digestion," pp. 524–525.
[53] Pagel, "Van Helmont, Gastric Digestion," pp. 525–527.
[54] Van Helmont, *Oriatrike*, p. 115.
[55] Boas, *Robert Boyle*, p. 58.

A fermentation occurring in the left ventricle of the heart was proposed by van Helmont as the source of the heat of the animal body. Respiration, far from serving to refrigerate the heart (as tradition had it), has as its major purpose the maintenance of the ferment and consequently the warmth of the body.[56]

Van Helmont's importance lies not as much in the specific reaction to which he attributed the production of the animal heat, but rather in the fact that he sought a specific cause (analogous to those found in the nonliving world) and refused to consider the warmth of the animal as an inherent quality. Once this move was made and animal heat was dethroned from its position as the major vital factor and attributed to a cause, fermentation, similar to others in nature, each physiologist was free to examine the phenomenon and develop a theory of his own to account for it.

Among the iatrochemists several adopted and attempted to extend van Helmont's theory of the physiological role of acids and alkalies. Franciscus Sylvius (François de la Boë, 1614–1672), a prominent student of van Helmont, developed the theory of ferments into a medical doctrine so broadly conceived as to explain with it many of the functions previously attributed to the four humors. Of particular interest is his proposal that the animal heat is produced by an active acid-alkali fermentation in the right side of the heart. This was brought about by the mixture of chyle- and lymph-filled blood from the upper great veins with bilious blood of the *vena cava*. The vital heat arose "on account of the different or rather opposite disposition

---

[56] John R. Partington, "Joan Baptista van Helmont," *Annals of Science*, 1 (1936), 375.

of each [kind of blood] in certain of their parts provoking an effervescence of great moment."[57]

Respiration, thought Sylvius, serves to temper the heat of the blood engendered in the heart, owing to the presence in the respired air of "nitrous and subacid particles able to condense the rarefied and boiling blood and so to gently restrain its ebullition."[58] It would be easy to be critical of this last observation of Sylvius' and see in it a continued adherence to the Galenic doctrine of the refrigerating action of respiration. On the other hand it is only our modern knowledge which allows such criticism, for to the careful observer the ambient atmosphere is almost always cooler than the body. The burden of proof would seem to lie with the man who suggests that the air serves a function other than cooling the lungs. In the early seventeenth century it was different to suggest on the one hand that the air imparts "vital spirits" to the body and on the other hand to propose that some "part" of the atmospheric air was necessary for the maintenance of a vital fire in the heart.

Some physiologists who were willing to recognize the cooling function of breathing were at the same time unwilling to admit that this was the principal use of respiration. But during this period of chemical solutions for

[57] Quoted by Sir Michael Foster, *Lectures on the History of Physiology During the Sixteenth, Seventeenth and Eighteenth Centuries* (Cambridge, Eng., 1924), p. 156. See also John M. Stillman, *The Story of Early Chemistry* (New York, 1924), p. 390.

[58] Quoted in Foster, *Lectures*, p. 157. See Franciscus Sylvius, *Opera medica* (Geneva, 1681), pp. 19f. Leonard G. Wilson, "The Transformation of Ancient Concepts of Respiration in the Seventeenth Century," *Isis*, 51 (1960), 168–170, notes that among early seventeenth-century chemists and physiologists the belief was common that nitrous particles were derived from niter or saltpeter and were cool in nature.

physiological problems there was little room left for "spirits." Nathaniel Henshaw was quite emphatic in his judgment: "As for the airs becoming the matter of Vital Spirits in Respiration, I shall say no more, than that I neither find any need of it, nor any way for the mingling of it, with the Mass of blood." In a variation on the Helmontian view, Henshaw believed that the motion of the lungs actually increased the fermentation and ebullition taking place in the blood. The great amount of heat which he thought was released in the left ventricle was compared with that which would come from the mixing of several chemical liquors. The heat which occurs in ferments of this nature is due to small particles of air, present in the liquor, which become dilated. Dilations of this sort, Henshaw asserted, are always accompanied by heat. An epitome of the mid-seventeenth-century chemists' view of life is offered by Henshaw: "To make short: Life itself is but a continuation of this vigorous fermentation of the blood, which is so long maintained, as the Mass of blood is kept hot, and circulating through the Veins and Arteries."[59]

Thomas Willis, a prominent English exponent of the iatrochemical approach, also believed that the animal heat arose from a fermentation, but was one of the first to adopt the view that fermentation itself "is an intestine motion of Particles, or the Principles of every body."[60] In a later treatise on the *Accension of the Blood* Willis joined his

[59] Nathaniel Henshaw, *Aero-Chalinos: Or, a Register for the Air* (1st ed., Dublin, 1664, 2nd ed., London, 1677), pp. 77, 86–87, 20, 94.
[60] Thomas Willis, *A Medical-Philosophical Discourse of Fermentation, or, of the Intestine Motion of Particles in Every Body*, trans. into English (London, 1681), p. 8. From the original Latin text, *Diatribae duae medico-philosophicae, quarum prior agit de fermentatione sive de motu intestino particularum in quovis corpore, altera de febribus sive de motu earundum in sanguine animalium* (London, 1659).

Oxford colleagues in explaining heat production by the interaction of the nitrous air with saline-sulphureous particles.[61]

Willis appears to have adopted a theory of fermentation which embodied several of the then current notions of chemical mixtures. Many of the elements of the Helmontian acid-alkali theory of ferments were present, but superimposed upon these was the notion that all fermentative liquors, blood included, are composed of heterogeneous particles of diverse figures and energies.[62] In this latter view Willis was surely influenced by the particulate theories of matter which had become quite prominent by the mid-seventeenth century.[63] The activity of the particles in the mixture could be increased or "unlocked by the adding of Ferments." In this case the native particles are freed from their "bonds" and induce a fermentation with "a more rapid motion and heat." This action Willis likened to a familiar chemical experiment in which fluid salts are mixed with a saline liquor of another kind, causing great heat and ebullition.[64]

It is exactly this type of fermentation "which the Blood suffers in the Bosom of the Heart" which brings the blood to great heat and carries it through the vessels of the

[61] This work appears in English translation as the fourth treatise of *Dr. Willis's Practice of Physick, Being the Whole Works of that Renowned and Famous Physician* (London, 1684), pp. 20–40; it appeared originally as *Dissertatione de sanguinis accensione, et de motu musculari* (London, 1670).

[62] Thomas Willis, *Of Feavers* (London, 1681), p. 48. This is an English translation of the second of the two medico-philosophical *diatribae* of 1659. In the English edition the pagination is continuous with the *Discourse of Fermentation*.

[63] For an excellent discussion of the development of theories of the particulate form of matter, see Marie Boas, "The Establishment of the Mechanical Philosophy," *Osiris*, 10 (1952), 412–541.

[64] Willis, *Of Feavers*, p. 48.

body.[65] If this phenomenon is to be understood, Willis suggested that we must determine the types of particles contained in the blood which make it fit to ferment. Relying on what is surely tradition, Willis concluded that the blood contains "very much Water and Spirit, a mean of Salt and Sulphur, and a little of Earth." The spirits are described as particles "always expansed, and endeavoring to fly away." The sulphur in the blood comes from the fat and sulphureous aliments taken in by the animal. It is the spirits together with the sulphureous particles that when passing through the heart, "being somewhat loosened, and as it were inkindled into a flame, leap forth, and are much expanded, and from thence they impart by their deflagration, a heat to the whole," upon which "the lively and vital heat in us depends."[66]

Although it seems clear that Willis thought of fermentations as the generalized chemical action upon which all other bodily processes depended, he hesitated when it came to the production of body heat. He could not ignore the Cartesian notion of "as it were a fire . . . set in the Chimny of the Heart." But, since Willis believed that fermentation involved the motion of particles, he was able to conclude that

there is little difference, whether the expansion of the Particles of the Blood, and exertion into the liberty of motion, be said to be done by Accension, or by Fermentation, forasmuch as by either way, the frame of the Blood may be so unlocked, that from thence the Particles of Spirit, Salt, and especially of Sulphur, being incited into motion, (as it were by an inkindled fire) may impart heat to the whole Body.[67]

[65] Willis, *Fermentation*, pp. 11–12.
[66] Willis, *Of Feavers*, pp. 48–49.
[67] *Ibid.*, p. 54. Accension refers to a boiling or distillation of fluids. See the *OED*.

When Willis returned to a discussion of animal heat in his treatise on the *Accension of the Blood*, he made one substitution in his earlier scheme; he dropped the talk of "spirit" and described instead "nitrous little bodies" which fill the air and "are everywhere ready for the constituting of fire and flame."[68] He had read the papers of Boyle and was almost certainly familiar with his work at first hand. There is every reason to assume that he was aware of the work of Hooke, Mayow, and Lower as well, the last being one of his students.[69] Willis adopted a theory of combustion similar to that of his Oxford colleagues and applied it in explanation of physiological heat.

Just as fire needs the nitrous particles of the air and a copious supply of sulphureous food, so too the flame in the blood requires both. In a manner very similar to fire, the vital flame emits hot effluvia full of soot and vapor. Any doubt that their nature is fiery is allayed by reflecting upon "the frequent burning of the mouth and tongue, and infecting them with blackness like the soot or smoke of a Chimney."[70]

Willis, much more influenced by the tradition of iatrochemistry than that of the new experimental chemistry of Boyle, Hooke, *et al.*, attempted to draw a very close and somewhat artificial analogy between the vital flame and the burning of a candle or lamp. He describes an imaginary machine which is essentially a large collection of oil lamps so constructed of tubes and pipes as to provide a constant

[68] Willis, *Accension of the Blood* (see note 61), p. 23. Willis readily admits (p. 25) that although he provided an explanation of fire or flame in his treatise on fermentation "we omitted there, that the accession of nitrous food was necessary for . . . sustaining it."

[69] *Ibid.*, p. 24. See R. T. Gunther, *Early Medical and Biological Science, Extracted from 'Early Science in Oxford'* (London, 1926), p. 64.

[70] Willis, *Accension of the Blood*, p. 24.

and perpetual heat in all its parts. "Indeed," Willis remarks, "such a Bannian or Bathing Engine, artificially made, might aptly represent the real Divine handy-work of the Circulation of Blood, and what burns in it, the Life-Lamp."[71]

In answer to the question of whether the blood itself is inflammable Willis reports that chemical analysis shows many particles of sulphur and spirit and a plentiful supply of inflammable oil. Luckily, it would seem, these are mixed with thicker elements "in a just proportion, to bridle their too great inkindling." The likeness between the vital flame and common fire is so great that Willis felt it necessary to explain why the former is never seen. Since the vital flame is enclosed in the heart and vessels, Willis conjectures that it might remain in the form of smoke, vapor, or breath. Even if it were to flame out, its shining might be so thin as to not be perceived by our sight just as, in the clear light of day, we are unable to behold a glowing red-hot iron.[72] Willis has gone beyond the data that the experimental study of combustion was yielding, as he had with fermentation, in his effort to provide a chemical explanation of the body heat which arose without visible fire or flame.

Among the various specific, mechanical causes which were chosen to account for animal heat, the newly revived atoms played a small but important role. Walter Charleton believed that spherical atoms of heat when moving with the proper velocity and direction served to vivify and warm the bodies of animals. Death is caused when atoms of cold insinuate themselves into the body in great swarms, overpowering the atoms of heat by impeding and suppressing their motions. The calorific atoms are then "no longer able

[71] *Ibid.*, p. 26.
[72] *Ibid.*, pp. 26–27.

to calefy the principal seat of life, the Vital flame is soon extinguished, and the whole Body resigned to the tyranny of Cold."[73]

What is significant about all the proposed alternatives to the innate heat is that they relied little, if at all, upon direct, new, experimental evidence. They represent a change in frame of hypothesis, rather than a significant augmentation of data. The heat of the animal body was elusive and the first attempts to ascribe a cause to it were based primarily upon analogy, at first to those chemical reactions in which heat was released without flame and only later to combustion itself.

The development of the conceptual framework within which animal heat, respiration, and combustion were linked can be attributed, for our purposes, to Robert Boyle, Robert Hooke, and John Mayow.[74] During the last four decades of the seventeenth century these men, and others within their circle in England,[75] carried out a number of remark-

---

[73] Walter Charleton, *Physiologia Epicuro-Gassendo-Charltoniana: Or a Fabrick of Science Natural, Upon the Hypothesis of Atoms, Founded by Epicurus, Repaired by Petrus Gassendus, Augmented by Walter Charleton* (London, 1654), pp. 314–315.

[74] I realize that I step here upon dangerous, well-trodden, if still confusing ground. The several parts which when brought together form the conceptual scheme mentioned have each been studied. I have benefited from all these studies but wish to mention two in particular: Henry Guerlac, "Studies on the Chemistry of John Mayow, I: John Mayow and the Aerial Nitre," *Actes du Septième Congrès International d'Histoire des Sciences* (Jerusalem, 1953), pp. 332–349; "II: The Poet's Nitre," *Isis*, 45 (1954), 243–255; and Douglas McKie, "Fire and the Flamma Vitalis: Boyle, Hooke and Mayow," *Science, Medicine and History*, ed. E. A. Underwood (London: Oxford University Press, 1953), I, 469–488. Guerlac has made amply clear that "nitre" was very literally in the air of the late seventeenth century. He has provided a detailed study of the origins of the idea of "aerial nitre" and I will not repeat his efforts.

[75] Guerlac's study of John Mayow has made clear who some of the leading figures were. Marie Boas, *Robert Boyle*, p. 185f, has suggested that "it was because the Royal Society acquired a vested interest in air that so much research . . . was carried out under its aegis." She suggests

able experiments which led them (or perhaps more accurately their initial theory led them) to conclude that atmospheric air, or some part of it, was involved in a similar manner in both combustion and respiration. In the latter process, as in the former, heat was produced. Thus they supposed that respiration did not cool the lungs but rather that breathing was the very process responsible for providing the vital heat.

Robert Boyle was led to his speculations about combustion and respiration through his work with the "pneumatical engine" or air pump which had recently been invented. In an early series of experiments carried out in the exhausted receiver of his pump Boyle confirmed the traditional suspicion that animal life and flame shared in a need for atmospheric air.[76] But Boyle would not accept that part of the tradition which asserted that the respired air was useful primarily to refrigerate the overheated heart or blood. Frogs, he pointed out, and other creatures without sensible heat had need to respire, as did old men whose natural heat is almost extinguished. The suggestion which was most attractive to Boyle in this paper of December, 1659 was a variant on the Galenic notion that the respired air serves to ventilate the blood and in some way aid in the disposal of "excrementitious vapours." But this proposal was not wholly acceptable, since Boyle had noted that the life of an animal left in a vessel from which the air had been

---

that with so much work involving the air pump and the physical nature of air it was inevitable that attention should be turned to the composition and structure of the air and that speculations should be made about its possible role in combustion and respiration, where it was known to be indispensable.

[76] Robert Boyle, *New Experiments Physico-Mechanical, Touching the Spring of the Air, and its Effects; Made, for the Most Part, in a New Pneumatical Engine* (1659), *Works of the Honourable Robert Boyle*, ed. Thomas Birch, vol. I (London, 1744).

pumped was shortened more than if the air had remained in the receiver even though there was less space left empty to receive the fuliginous steams.[77] Respiration remained, for Boyle at this period, more rather than less mysterious than when his experiments were started.[78]

With little more gained than an intuitive feeling that respired air did something other than just receive the waste thrown out of the blood in its passage through the lungs, Boyle left his studies of respiration and did not return to them until 1672.[79] The problem of the relation of respiration to vital heat did, however, capture the interest of others. Henry Oldenburg, writing to Boyle in 1664, communicates some of the optimism engendered by current experiments and observations in London and Oxford. He wondered if "we shall at length find out more for the use of respiration, and the account upon which it is so absolutely necessary." If he is communicating current opinion, Oldenburg's further questions indicate that the problem was still far from solved. He asks, for instance, if it will "not be made out at last, that Life is a kind of subtil and fine flame? to which the air must be applied, to keep both it in motion, and the blood, where in it resideth."[80]

Robert Hooke, who had served as an assistant to Boyle and had constructed the vacuum pump for him, began a series of experiments upon respiration in the winter of 1662/3. On January 28, "Mr. Hooke made the experiment of shutting up in an oblong glass a burning lamp and a chick; and the lamp went out within two minutes, the

[77] *Ibid.*, pp. 66–67.

[78] McKie, "Flamma Vitalis," pp. 470–471.

[79] John F. Fulton, "Robert Boyle and His Influence on Thought in the Seventeenth Century," *Isis*, 18 (1932), 88–90.

[80] Oldenburg to Boyle, Nov. 17, 1664, Boyle, *Works*, V, 322–323.

chick remaining alive, and lively enough."[81] The hopes for a simple solution to the relation between combustion and respiration received a setback, for a difference rather than a similarity between the two processes was implied.

Hooke returned to the study of respiration in 1667 with a more rewarding series of experiments, in which he was able to demonstrate that the important feature of breathing was the supply of fresh air it provided and not the mechanical action of the organs involved. This was accomplished by opening the thorax of a dog, removing the diaphragm, and keeping the lungs supplied with fresh air by means of a bellows.[82]

It was in response to one of these experiments that Boyle related the facts about a man who was able to remain under water for three hours without showing any signs of harm. The members present at the Royal Society meeting that day then entered a discussion about "what quality it was, that made the air fit for respiration." The traditional view about the vapors with which the air became clogged and entangled, thus making it unfit, was proposed by some members, but Hooke expressed the opinion "that there is a kind of nitrous quality in the air, which makes the refreshment necessary to life, which being spent or entangled, the air becomes unfit."[83] During the same session Hooke reminded his colleagues of an experiment which demonstrated that a coal fire burning in an enclosed box (within which the air was circulated) rendered the air unfit to support fresh

[81] Thomas Birch, *The History of the Royal Society of London* (London, 1756), I, 180.

[82] R. T. Gunther, *The Life and Work of Robert Hooke, Early Science in Oxford*, VI (Oxford, 1930), pp. 305, 309, 310, 315, 317. These experiments were continued into 1668, see pp. 331ff.

[83] *Ibid.*, p. 309, June 27, 1667.

fire. Hooke in this discussion had raised two points, both important in the conceptual scheme which was being developed, but neither of them wholly new or attributable to Hooke himself.

The similarity of the effects upon the air of respiration and combustion had been noticed, and unsuccessfully accounted for, by many others. The suggestion that atmospheric air contains a nitrous quality which renders it fit for the support of respiration and combustion had a long history, with one of its most important early proponents in England being Dr. George Ent.[84] Ent, who was still attending meetings of the Royal Society during the period of Hooke's experiments,[85] had proposed, in his *Apologia pro Circulatione Sanguinis* (1641), that a nitrous substance was absorbed by the blood during respiration and served to maintain the fire in the heart in a manner similar to ordinary combustion.[86] Hooke himself, in the *Micrographia* (1665), had proposed a theory of combustion in which he supposed that the air was composed of parts which differed in their behavior, one of these parts being active in combustion.[87]

Although Hooke's theory of combustion did not seem to gain wide acceptance among his associates,[88] others, working on similar problems, presented theories of respiration and combustion conceptually identical to Hooke's. The *Philosophical Transactions* of 1668 carries a review of John Mayow's *De Respiratione* in which the Oxford physiolo-

[84] Guerlac, "John Mayow," *Actes de Septième Congrès*, pp. 341–342.

[85] Gunther, *Robert Hooke*, p. 315, records that Sir George Ent was present and discussing Hooke's respiration experiment on Oct. 10, 1667.

[86] Guerlac, "John Mayow," *Actes du Septième Congrès*, p. 342.

[87] Robert Hooke, *Micrographia* (1665), *Early Science in Oxford*, ed. R. T. Gunther, XIII (Oxford, 1938), pp. 103–105.

[88] McKie, "Flamma Vitalis," p. 476, suggests that during the five years following its publication Hooke's theory of combustion did not appear to gain many followers in the Royal Society.

gist's proposed theory of respiration is spelled out at length. Mayow rejected those opinions that "would have Respiration serve either to cool the heart, or to make the Bloud pass through the Lungs out of the right ventricle of the heart into the left, or to reduce the thicker venal Bloud into thinner and finer parts." Instead he proposed that in breathing something absolutely necessary to life was conveyed from the air to the blood, "which, whatever it be, being exhausted, the rest of the Air is made useless, and no more fit for Respiration."[89]

In that portion of the air important for the maintenance of life there exists, Mayow suggested, "the more *subtile* and *nitrous* particles." This "aerial nitre" was seen as "necessary to *all* life . . . the plants themselves do not grow in that Earth, that is deprived thereof." The nitre, Mayow believed, became mixed with the sulphureous parts of the blood, causing a fermentation to take place. This ferment was found not only in the heart, but in the blood in the pulmonary vessels themselves, and also in the arteries. This constituent of the respired air was viewed as also being "highly necessary to the motion of the heart, for as much as the heart is one of the Muscles, the motion of every one of which absolutely requires this *Aerial Nitre*."[90] This view of nitrous particles, published by Mayow in 1668, is strikingly similar to the one verbally reported by Hooke in 1667. Thus by 1668 the hypothesis that something contained in the air was necessary for respiration and that it behaved in the process in a manner similar to that which had already been described for combustion (by Hooke in

---

[89] "An Account of Two Books, 1. Tractatus Duo, Prior de Respiratione; Alter de Rachitide, A. Joh. Mayow, Oxon, 1668," *Phil. Trans. Roy. Soc.* (*London*), 3, No. 41 (1668), 833–835. This 1668 review would indicate that Mayow's views were immediately and widely known.

[90] *Ibid.*, p. 833.

the *Micrographia*) was a part of the scientific literature.

Mayow's contemporary at Oxford, the anatomist-physiologist Richard Lower, provided another kind of evidence for the "nitrous air" hypothesis of respiration. In his famous *Tractatus de Corde* (1669) Lower was at pains to put an end to those theories (probably adopted from Descartes) which suggest that the "Heart contains a sort of vestal Fire, which so heats the inflowing blood that it must immediately pass on out into the arteries," thus accounting for the circulation. The movement of the blood, he contended, is independent of any heating it may receive in the heart. Lower is not at all sure that it is true to say that heat is produced only in this organ, or that the blood is warmed only by it. He finds nothing in the heart sufficient to produce so much heat, and, even more important to Lower, he was able to report that "if we insert the fingers into the Heart of an animal during a vivisection experiment, we do not feel a heat of this intensity" and, furthermore, "the pericardial fat could not solidify, as it does, under such circumstances." But, having provided direct refutation of the earlier claims of Fernel and Descartes, Lower is forced to admit that "we are certainly warmed by a fire that is more than fictitious or metaphorical."[91]

Lower himself provided no explanation of how the body is warmed but instead refers the reader to a forthcoming treatise on *The Spirit, and Also the Heating of the Blood* by the learned Dr. Willis.[92] Having removed the site of heat production from the heart to the blood itself, Lower

[91] Richard Lower, *Tractatus de Corde item de Motu & Colore Sanguinis et Chyli in eum Transitu* (London, 1669), trans. K. J. Franklin, *Early Science in Oxford*, ed. R. T. Gunther, IX (Oxford, 1932), p. 70; p. 72, he cites Lambert Velthuysen (b. Utrecht 1622, d. 1685); p. 74.

[92] *Ibid.*, p. 74. See discussion above. Lower is almost surely referring to the *Dissertatione de sanguinis accensione, et de motu musculari* (1670), published one year later than the *Tractatus de Corde*.

proceeded to provide one piece of sound experimental evidence for the conjecture that the blood in passing through the lungs was altered by the respired air it encountered in them. Lower was able to demonstrate that the color difference between venous and arterial blood was generated in the lungs and occurred quite independently of any heating of the blood in the heart. He attributed the red color of arterial blood in one paragraph to the penetration of particles of air into the blood, and in another to "nitrous spirit of the air" which is mixed with the blood.[93]

But lest it seem that all the experiments that were being carried out by the members of the Royal Society pointed in the direction of the theory which we know to have emerged, it is important to view another set of experiments reported by Robert Boyle in 1670 and 1671. He enclosed a mouse together with a mercury gauge in a sealed vessel. After two hours, during which the mouse continued to breathe, there was no apparent change in the pressure of the air, from which Boyle concluded that "Air, become unfit for Respiration, may retain its wonted pressure."[94] In another experiment Boyle took a bird and a burning candle and sealed them together under a glass receiver. Each time this experience was repeated, the bird outlived the flame. These results led Boyle to question whether the "Common flame" and the Vital Flame are maintained by distinct substances or parts of the air or, alternatively, whether the one constituent of the air which nourished both was used up more rapidly by the common flame while there remained enough for the more temperate vital flame to keep alive.[95]

His findings in these two sets of experiments led Boyle

[93] Lower, *Tractatus de Corde*, pp. 164, 168, 169.
[94] Robert Boyle, *Phil. Trans. Roy. Soc. (London)*, 5 (1670), 2046–2047.
[95] Robert Boyle, *New Experiments About the Relation Betwixt Air and the Flamma Vitalis of Animals* (1672), *Works*, III, 261–262.

to remain much more cautious than his compatriots when
he outlined his theory of respiration and combustion in
1674. He willingly conceded that his observations of the
necessity of air for the maintenance of flame and fire made
him "prone to suspect, that there may be dispers'd through
the rest of the Atmosphere some odd substance, either of a
Solar, or Astral, or some other exotic, nature, on whose
account the Air is so necessary to the subsistence of Flame."
But, whatever it is that supports the combustion, it is an
imperceptibly small bulk of the air it impregnates. For
upon extinction of the flame the *elasticity* (volume) of the
air is not diminished. Furthermore, Boyle continues,

this *undestroy'd springyness* of the Air seems to make the necessity of fresh Air to the Life of *hot* animals, (few of which, as
far as I can guess after many tryals, would be able to live two
minutes of an hour, if they were totally and all at once deprived
of Air,) suggest a great suspicion of some *vital substance*, if I may
so call it, diffus'd through the Air, whether it be a *volatile Nitre*,
or (rather) some *yet anonymous* substance, Sydereal or Subter-
raneal, but not improbably of kin to that, which I lately noted to
be so necessary to the maintenance of other flames.[96]

Thus Boyle, in a tentative manner, has developed a unified
explanation of the action of air in the processes of respiration and combustion.

Robert Hooke, in his formulation of the relation between
respiration, combustion, and vital heat, was somewhat less
cautious. Fresh air he believed to be "the Life of the Fire,
and without a Constant supply of that it will go out and
Die." Analogous to this is the "Life of Animals, who live
no longer than they have a constant supply of fresh Air to

[96] Robert Boyle, *Tracts: Containing I. Suspicions About Some Hidden
Qualities of the Air; with an Appendix Touching Celestial Magnets, and
Some other Particulars* (London, 1674), pp. 24, 26–27.

breath, and, as it were, blow the Fire of Life." Turning to the question of how heat is actually generated within the animal, Hooke describes a process which is almost identical to the one he had already proposed for combustion. He suggested that "the Heat in Animals . . . [is] . . . caus'd by the continual working of the Liquors and Juices of the Body one upon another, and more especially by the uniting of the Volatile Salt of the Air with the Blood in the Lungs." The process was likened to "a kind of Corrosion or Fermentation."[97]

In his *Tractatus Quinque* of 1674 John Mayow enlarged upon the theory of respiration that he had presented in 1668 and extended the general framework outlined for respiration to combustion, and a number of other physiological functions.[98] The similarity of his theory of aerial nitre to the "nitrous spirit" or "volatile salt" of Robert Hooke is abundantly clear;[99] the extent to which Mayow has developed his observations and hypothesis into an integrated conceptual system lends to his work an added significance.[100]

Mayow carried out a series of experiments, on respiration and combustion, which were conceived in a manner similar

[97] Robert Hooke, *The Posthumous Works of Robert Hooke, containing the Cutlerian Lectures and Other Discourses*, ed. Richard Waller (London, 1705), pp. 111, 50.

[98] I have utilized John Mayow, *Medico-Physical Works, Being a Translation of "Tractatus Quinque Medico-Physici (1674)"* (Alembic Club Reprints, No. 17, 2nd ed.; Edinburgh: Alembic Club, 1957).

[99] Guerlac, "John Mayow," *Actes du Septième Congrès*, p. 336.

[100] A glance at the table of contents provided in the Alembic Club edition of the *Tractatus Quinque* gives evidence of the range of phenomena which Mayow hoped to explain on the basis of the "nitro-aerial spirits." The four tracts of immediate interest are *De Respiratione* (of which an earlier edition was published in 1668), *De Sal-Nitro et Spiritu Nitro-aero*, *De Respiratione Foetus in Utero et Ovo*, and *De Motu Musculari et Spiritibus Animalibus*.

to those which had been done by Boyle. A difference in experimental technique, however, caused different results which in turn permitted Mayow to offer a stronger hypothesis concerning the relation between breath and flame. Rather than experimenting, as had Boyle, in the receiver of a vacuum pump or a sealed vessel within which a mercury barometer had been enclosed, Mayow utilized a glass vessel inverted over water.[101] Because of what we now know to be the release of carbon dioxide in both respiration and combustion, and the dissolving of this gas in water, Mayow was able to observe a reduction in the volume of enclosed air due to breathing and fire.[102] The conclusion that Mayow was able to draw was just the one upon which Boyle had hesitated. Mayow had demonstrated that "air is deprived of its elastic force by the breathing of animals very much in the same way as by the burning of flame" and from this went on to assert that "we must believe that animals and fire draw particles of the *same kind* from the air." This last, important assertion had its basis in comparative experiments. Mayow claims to have found "that an animal enclosed in a glass vessel along with a lamp will not breathe much longer than half the time it would otherwise have lived."[103] He was confident that in observations of this sort he had

[101] Mayow, *Medico-Physical Works*, pp. 72–73. He also utilized a moistened bladder stretched over the opening of a glass vessel. For illustrations of the apparatus, see Table 5 at end of the volume.

[102] *Ibid.*, p. 73. "I have ascertained from experiments with various animals that the air is reduced in volume by about one-fourteenth by the breathing of the animals." Thomas Beddoes, in his *Chemical Experiments and Opinions Extracted from a Work Published in the Last Century* (Oxford, 1790), p. xix, claims that the experiments of Torricelli were a more important influence upon Mayow than were the works of Boyle. Beddoes would thus attempt to explain Mayow's success in this area where Boyle failed.

[103] Mayow, *Medico-Physical Works*, p. 75. Italics mine.

found experimental verification for the analogy drawn between combustion and respiration.

The mechanism which Mayow invoked in order to explain these two important processes resembles the one proposed by Hooke and shares with it the same difficulties. Perhaps the major problem faced by both of these seventeenth-century scientists was the fact that they had to deal indirectly with the part of the air that they believed to be active. The inability to isolate and identify the "nitrous air," or "nitro-aerial particles (spirit)" left their fine proposals in the state of unverified hypotheses. The conceptual (as distinguished from the experimental) framework within which they worked, however, *was anticipatory* of developments made in later generations. This is particularly true of the physiological explanations offered by Mayow.

As an animal breathes, according to Mayow, air is taken into the blood and there deprived of its nitro-aerial particles, in consequence of which when expired the air is seen to have lost its elasticity.[104] Within the blood the nitro-aerial particles are thoroughly mixed with the saline-sulphureous particles found there, thus causing "a very marked fermentation," and with it the effervescence responsible for animal heat. That an effervescence of this nature is capable of producing heat is demonstrated by the exposure of any saline-sulphureous minerals (such as newly dug up vitriolic marcasites) to moist air which causes an effervescence and then intense heat. Combustion had also been viewed as being caused by the mixture of the nitro-aerial and saline-

[104] *Ibid.*, pp. 94, 204. This view is proposed in preference to an earlier one in which the lungs were seen as the site of separation. A similar difficulty arose later when the site of oxygen separation was being considered.

sulphureous particles. However, Mayow comments, in criticism of past proposals and in anticipation of future arguments, "although flame and life are sustained by the same particles it is not on that account to be supposed that the mass of the blood is really on fire."[105]

In the development of his theory of respiration and combustion, Mayow was quite consciously constructing a rational explanation in order to avoid the necessity of having "recourse to an imaginary Vital Flame that by its continual burning warms the mass of the blood." Descartes's scheme, in which the beating of the heart was seen as the result of a process of heating and rarifying the blood, is specifically criticized by Mayow.[106] In a statement very similar in tone to one by Harvey, Mayow points out that the heart contracts because it is muscle and owing to its continuous labor it causes the active mixing of nitro-aerial and sulphureous particles from which "a notable heat must be produced." The heat which arises during all muscular activity or violent motion is ultimately reduced to a greater effervescence of the particles of the air and blood. The rapid breathing which accompanies such movement is not used, as some had supposed, for the cooling of the heated blood but rather is responsible for just the opposite effect as it replaces the nitro-aerial particles used up in the heightened fermentation. Mayow particularly criticizes the notion which attributes the heating to the motion of the body itself; "there is no such friction of the parts (from which alone heat arises) as could account for so intense a fervour."[107]

[105] *Ibid.*, pp. 102, 104, 77.
[106] *Ibid.*, pp. 108, 206.
[107] *Ibid.*, pp. 249, 245. The idea here specifically criticized by Mayow became the basis of the theory of animal heat proposed by the iatro-mechanics of the early eighteenth century. See Chapter IV.

The reason we have devoted so much space to these late-seventeenth-century theories of animal heat is that they represent the period of major shift from the doctrine of an "innate heat." Not only did Boyle, Hooke, Mayow, and their contemporaries deny the existence of a warmth inherent in the animal body and propose in its place a source of heat analogous to one in inanimate nature, but they went further and attempted to find experimental verification for the proposed analogy. Van Helmont had tried to reproduce the process of digestion (acid fermentation) *in vitro* and failed.[108] Boyle, Hooke, and Mayow had somewhat greater success in their attempts to find an *in vivo* phenomenon, respiration, reproduced *in vitro*, combustion.

Historians of chemistry have pointed out that the aerial-nitre theory of combustion did not survive the end of the seventeenth century despite its mention in "that bible of 18th-century experimental science," the *Opticks* of Isaac Newton.[109] The physiological counterpart, the theory of respiration, was also replaced by a new eighteenth-century proposal, different from the one found in chemistry, and interestingly enough having the mechanical theories of heat originated by Isaac Newton, Robert Boyle, and others as its basis. The new theory of animal heat was in no way better founded in experimentation but seemed instead to represent a change in orientation upon the part of the physiologist; he no longer looked toward chemistry for his analogy but turned instead to physics. Some measure of the confused legacy of which the eighteenth-century scientist was the recipient is found in the entry under "Heat" in John Harris's *Lexicon Technicum*. Heat, which

[108] Boas, *Robert Boyle*, p. 58.
[109] See Guerlac, "The Poet's Nitre," p. 255. For other minor references to Mayow in the eighteenth century see J. R. Partington, "The Life and Work of John Mayow (1641–1679)," *Isis*, 47 (1956), 222–223.

is considered one of the four primary qualities, seems to consist only, or at least *chiefly*, in the local Motion of the small Parts of a Body Mechanically modified by certain Conditions, of which the Principle is the vehement and various Agitations of those small Insensible Parts." In the same volume the author turns to the *flamma vitalis* which is described as "a fine and kindled, but mild substance . . . a *Vital Flame*," and to whose preservation the "Air taken in by Respiration, [is considered] to be necessary, as it is to the Conservation of ordinary Flame." The authority cited for this view is Robert Boyle who "found that the Vital Flame of Animals (if Life may be so called) did survive or outlast the Flame of Spirit of Wine, or of a Wax, or Tallow Candle."[110]

One important contribution to the study of animal heat made during the seventeenth century was the invention and development of the thermometer. The problem of determining relative heat intensity is immediately apparent, and the thermometer has proved an indispensable tool in the study of animal heat. Partington claims that van Helmont had developed an air thermoscope with which he measured the temperature of the body.[111] The question which most exercised the physiologists was whether or not the heart and blood through a high degree of heat gave an indication of a vital heat in the body. Fernel had claimed to have found an excessive heat about the heart when he inserted his fingers into an opening in the chest cavity.[112] An overheated heart should be expected according to the theory of Descartes. Lower claimed that no such heat

---

[110] John Harris, *Lexicon Technicum: or, an Universal English Dictionary of Arts and Sciences: Explaining not only the Terms of Art, but the Arts Themselves* (London, 1704).

[111] Partington, "Joan Baptista van Helmont," p. 378.

[112] *Supra*, Chapter II.

existed about the heart, but he recorded no thermometric measurements.

Robert Boyle related the results of an inquiry into the heat of human blood. Using "a sealed weather-glass, adjusted by the standard of *Gresham* college" he had a surgeon put the instrument "into the porringer, wherein he was going to bleed a young gentlewoman." The liquor in the stem ascended to within about an inch of the smaller, upper ball of the thermoscope. However, in the case of a healthy middle-aged man "the tincted spirit of wine ascended above all the marks belonging to the stem." Boyle concluded that "the warmth, that made it rise, did considerably exceed the usual warmth of the air in the dog-days." Some indication of the scale of degrees is gained by Boyle's pointing out that the thermoscopes in use were gauged so that the liquid would remain within the stem all year long, "without sinking quite into the greater ball in winter, or ascending into the lesser in summer."[113] No speculation was offered as to why the blood of the healthy middle-aged man was considerably warmer than that of the young gentlewoman.

An important direct measurement, of the heat of the heart, was made by Giovanni Borelli. Like Lower, he was unable to discern any burning heat upon inserting his finger into an opening in the chest of a living animal. In order to determine exactly the degree of heat of the heart, Borelli utilized a thermometer of the type developed by members of the Accademia del Cimento. The instrument was placed within the left ventricle of the heart of a stag whose chest had been laid open. The greatest degree of heat of the heart did not exceed forty, which Borelli tells us is as much as the heat of the summer sun. Using a

[113] Robert Boyle, *Memoirs for the Natural History of Human Blood* (1684), *Works*, IV, 166.

similar thermometer Borelli then measured the heat of the liver, lungs, and intestines of the same living stag, and found these organs to have the same degree of warmth as the heart and other parts of the viscera. While lending weight to the argument against the heart as the seat of an innate warmth, Borelli's measurements could hardly be considered conclusive, for he would be unable to answer the critic who claimed that respiration was all the time cooling the heart. His measurements would prove damaging against an explanation of the type advanced by Descartes, one which called for a great boiling or rarefying of blood in the heart. Borelli had concluded to his own satisfaction, however, that no fire or actual flame could exist in the heart.[114]

The nature of physiological explanation had clearly changed during the course of the seventeenth century and the theory of animal heat directly reflected these alterations. At the beginning of the century it was possible to assume that an innate heat "burned" in the heart even while other physiological phenomena were beginning to yield to experimental analysis. By the close of the century not only was a causal explanation sought for the heat of the warm-blooded animal, but that explanation was almost identical to the one advanced in a general theory of combustion. Of almost equal importance was the move to verify, by instrument, the presence of the supposed special heat in the heart. Even though thermometers were in their infancy, and could measure only degree or intensity and not content of heat, the assumption was now made that a physical verification would indicate whether or not the heart was an extra warm source from which heat could flow to the rest of the body.

[114] Giovanni Borelli, *De Motu Animalium* (2nd ed.; Lugduni Batavis, 1685), II, 137–138.

·IV·

"Millstones in the Stomach"

Millstones were brought into the Stomach, Flint and
Steel into the Blood-vessels, Hammer and Vice into
the Lungs.

John Stevenson[1]

Just why the combustion-respiration theory of John
Mayow and his contemporaries did not survive the closing
years of the seventeenth century remains unclear. It cannot
really be argued that this was due to the experimental un-
verifiability of the theory, for the proposals which replaced
it were in no way better adapted to experimentation. This
is particularly true of the early-eighteenth-century hypotheses designed to account for the generation of animal heat.
The physiologists adopted the suggestion that the source of
heat was in some way to be found in motion, in this particular case the motion of the blood through the vessels of
the body.

This mechanical theory of body heat certainly had some
of its roots in the iatrophysical ideas traceable to Descartes
and Borelli. In large measure, however, iatrophysics had

---

[1] John Stevenson, "An Essay on the Cause of Animal Heat, and on
some of the Effects of Heat and Cold on our Bodies," *Medical Essays
and Observations* (4th ed., Edinburgh, 1752), V, pt. 2, pp. 326–413; 1st ed.
(1733–1744), p. 352.

already achieved its greatest successes in the works of men like Santorio and Baglivi and seemed to have run its course by the end of the seventeenth century. The persistence of iatrophysical theories well into the eighteenth century and their strength, which was particularly noticeable in Britain, was certainly due in large measure to the successes of Isaac Newton's mechanical philosophy. In many instances the iatrophysicists adopted the "Newtonian" techniques of argument and often adorned their works with his name.

One thing that the mechanical philosophers held in common was their rejection of the chemical theories of animal heat which had been so widely held in the latter half of the seventeenth century. Archibald Pitcairne, one of the more extreme, but also influential, British iatromechanists, outlined a theory of animal heat which maintained its prominence well into the eighteenth century.[2] In the clearest of language Pitcairne asserts that "by *Innate Heat*, ought therefore to be understood, that Attrition of the Parts of Blood, which is occasioned by its circulatory Motion, especially in the Arteries." The blood vessels are likened to a hollow cone, diminishing in size from a circular base to the apex. The blood, forced into the vessels by the heart, meets resistance from "the sides of the Arteries, and from the

[2] Archibald Pitcairne (1652–1713) is best known for his posthumously published lectures *Elementa Medicinae Physico-Mathematica, libris duobus quorum prior theoriam, posterior praxim exhibet; In medicinae studio sorum gratiam delineata* (London, 1717). There is also an edition at the Hague, 1778 and another at Leyden, 1737. I have examined an English translation of 1727 by George Sewell and J. T. Desaguliers, and another, *The Philosophical and Mathematical Elements of Physick . . .* etc., translated by John Quincy (2nd ed., London, 1745). (All citations are of the latter fairly accurate rendering.) The translator's prefaces in both English editions note Pitcairne's tie to the Italian mechanical tradition of Galileo, Torricelli, Borelli, and Bellini. Quincy credits Pitcairne with being "the first . . . on this side of the Alps acquainted with the true Method of Reasoning in the Art of Physick." (*Elements of Physick*, p. iv).

preceeding Blood." Heat arises because the blood contains in it parts "fitted to excite Heat." The amount of heat is considered to be proportional to the amount of attrition and abrasion undergone. The greater the velocity of the blood, the more frequent are the strokes against the sides of the arteries and the preceeding blood, therefore the heat will be increased.[3]

What Pitcairne has done is to extend the rules covering the behavior of inanimate bodies to living systems. This is in essence what the iatrochemists had done, with the difference that their fundamental theories were chemical in nature, whereas Pitcairne's are mechanical or physical. His approach to animal heat is that of a man very much impressed by the successes in the studies of bodies in motion. He set forth two proposals treating the heat of the blood:

first, that at the same Distances from the Heart, the Heat of equal Quantities of Blood will be as their Velocities. Secondly, In the same Velocities of Blood, the Heat will be reciprocally as the Distances from the Heart.

The argument is simple in the extreme. Effects are proportional to their causes; resistance is proportional to velocity, and the velocity is greatest closest to the heart. "The Heat of the Blood," Pitcairne claimed, "may be considered as a Rectangle under the Velocity and the Distance."[4] Body heat is thus reduced to a simple term varying in proportion to velocity and distance.

Having established, to his own satisfaction, a theory of animal heat, Pitcairne turned to examine the ideas he was attempting to replace, a heat thought to arise from fermentation of intestine motion in the blood. Although he dismissed the idea that a ferment in the heart could heat the

[3] *Ibid.*, pp. 20–21.
[4] *Ibid.*, pp. 21–23.

blood as impractical, since the ferment would be washed out with every contraction, Pitcairne presented arguments against the several theories of fermentation found popular by many of his contemporaries. Turning to the view held by "many eminent moderns" that the heat of the blood is due to a fermentation maintained by the action of a "subtile Matter," Pitcairne objected that the same cause should excite heat in blood not in the vessels, for instance, in a basin. Everyone knows, however, that "Blood out of the Vessels, loses all that Heat, and imaginary Fermentation."[5]

Pitcairne went on to suggest that in all likelihood there was no fermentation in the blood. A real fermentation, he reminded his readers, "is a mutual Action of an Alkali and an Acid upon one another with Ebullition." The action of such a process was believed to turn the oils of a liquid into an inflammable spirit. That the blood contains no inflammable spirit, but only a urinous one, is proof enough to Pitcairne that there is no real fermentation in the blood. Furthermore, he pointed out, if anyone will take the trouble to distill the blood of a healthy person he will find no acid present in the blood, thus removing one condition necessary to a true fermentation.[6]

Although he considered the foregoing argument against a fermentation in blood sufficient for those "who seek only the Truth," Pitcairne added still more for those who "contend merely about Words, and insist . . . that there is such a thing as Fermentation in the Blood, although there is no Acid in it." He referred specifically to Thomas Willis, who defined animal heat as arising from a fermentation or intestine motion of the constituent parts of the blood. Taking his readers through a discussion of homogeneous and non-

[5] *Ibid.,* pp. 24–26; *materiam subtilem.*
[6] *Ibid.,* pp. 28–30.

homogeneous fluids, Pitcairne led them to the conclusion that there could be no intestine motion of the parts of the blood and therefore no heat produced by such motion.[7]

Pitcairne's insistence that the animal heat arises from the attrition of the parts of the blood remains as nothing more than an assertion based upon deductions from general mechanical principles. For although he carried out, and reports, numerous experiments upon the blood, they involved nothing more than the mixing of blood with other substances.[8] There are no experiments upon blood, or any other liquid, which would indicate a source of Pitcairne's attrition theory of animal heat.

Hermann Boerhaave, one of the most influential physicians of the century, followed Pitcairne in adopting a mechanical explanation for the production of animal heat. In his *Elements of Chemistry*[9] he proposed an answer to the question "which has so much exercised the skill of Chemists, Physicians, and Philosophers; *viz.* whether the human Blood has the greatest degree of Heat in the Heart? And if this is the case, what is the reason of it?" He exclaimed at the "dissertations . . . we find amongst Authors upon this subject and what very different opinions about it!" Boerhaave claimed that the blood in the veins is the coldest. It circulates through the cold extremities of the body in a lax and weak manner, the veins themselves being inactive. But the cool blood which has collected in the right ventricle is pressed and driven by the contraction of the ventricle into the narrow, elastic, strong branches of the pulmonary artery. This contraction, aided by "the vast action of respira-

---

[7] *Ibid.*, pp. 30, 34.
[8] *Ibid.*, pp. 39ff.
[9] Hermann Boerhaave, *Elements of Chemistry: Being the Annual Lectures of Hermann Boerhaave, M. D.*, trans. Timothy Dallowe (London, 1735).

tion," forces as much blood through the lungs alone as is simultaneously passed through the rest of the body. The blood, therefore, "would in no part of the Body, suffer so great an attrition, and of consequence acquire so much Heat as in the Lungs alone."[10]

In this view Boerhaave has joined Georg Stahl in proposing that the friction engendered in the lungs is one of the chief sources of animal heat.[11] But the lungs, for Boerhaave, also served to moderate the heat produced therein by exposing the blood to the inspired air as it passes through "a vast number of exceeding fine Arteries, which are applied all around the thin vesicles of the Lungs." Thus the paradox that the blood "is cooled in no part of the Body more, in this respect, than it is in the Lungs." Boerhaave remarked on the surprising dual function of the lungs and proposed the following explanation:

> The Blood, and recent Chyle, could not be propelled through all the vital pipes of the whole machine without endangering the animal life, if it was not first vastly divided, and reduced into its most subtil Elements by the forcible attrition of the Lungs; but this could not be effected without a great production of Heat.[12]

And further, if the blood were not sufficiently cooled the animal would die of a pestilential disease. It was in an effort to investigate the cooling function of the air that Boerhaave asked his students Fahrenheit and Provost to determine experimentally how great a degree of heat of the ambient air

[10] *Ibid.*, vol. I, p. 162. The discussion of body heat in the edition of Peter Shaw (1741) is quite similar to this one. See pp. 292–294.

[11] Michael Foster, *Lectures in the History of Physiology During the Sixteenth, Seventeenth and Eighteenth Centuries* (Cambridge, Eng., 1924), p. 223. Pitcairne had not specified the lungs or any other part of the vascular system as being the site of greatest heat production. Velocity and proximity to the heart were the sole measure.

[12] Boerhaave, *Elements of Chemistry* (1735), pp. 162–163.

animals could endure.[13] The production of body heat was for Boerhaave a by-product of the preparation of the chyle or nutritive matter.

That Boerhaave believed heat could be produced by friction is quite clear. He earlier had explained that

it appears evident from modern Experiments, that a surprising Heat and Fire may be instantly produced in the coldest, hardest, and heaviest Bodies, purely by their attrition with the lightest, softest, cold Fluids, if the motion is exceeding violent.[14]

Boerhaave has here accepted the tradition of Boyle and Newton, but that this was not his total thought on the subject of heat can be seen in the pages of the *Elements of Chemistry*.[15] Hooke had attempted to distinguish mere heat from fire and flame. For him, animal heat was produced by a process analogous to combustion, but for Boerhaave it might be said that the heat of the animal body was not viewed as being similar to combustion but rather as analogous to that type of heat which arises from motion or agitation. Like Hooke, Boerhaave attempted to identify the sources of heat and in so doing developed a theory that varied from the Newtonian view. He proceeded to distinguish two types of fire, elementary fire, and the fire supported by combustion. The former he described as being corporeal and having the ability to enter the vacuities or pores of solid bodies. It would seem that Boerhaave believed that it is the substance of fire which is ultimately responsible

13 T. C. Allbutt claimed that this is the only physiological experiment that can be credited to Boerhaave. See F. H. Garrison, *An Introduction to the History of Medicine* (4th ed., Philadelphia, London, 1929), p. 316n.

14 Boerhaave, *Elements of Chemistry* (1735), p. 112.

15 See also Hélène Metzger, *Newton, Stahl, Boerhaave et la Doctrine Chimique* (Paris, 1930), pp. 209–245.

for the heat of the animal body. Several pages after he presented what appears to be a simple mechanical theory of animal heat he returned to the discussion in another context and talked of "a perpetual collection of Fire, and a communication of Heat, by that attrition of the parts, which is necessary to the support of animal life."[16]

Boerhaave, unlike Boyle, Hooke, and Mayow, was not at all concerned with any alterations in the air which was inspired by an animal; he was solely interested in the refrigerating effects of respiration. Why this shift in emphasis occurred is hard to say, except that it can be identified with the general move away from the chemistry of fermentation and coction.[17] On the strength of his views in several other areas of physiology Boerhaave should be identified with the iatromechanical tradition.[18] Fundamentally, he believed that all organic phenomena have their basis in the movement of fluid and solid parts of the body. If this is recognized, his acceptance of an attrition theory of animal heat comes as no surprise.

Boerhaave left no indication of whether he made direct measurements of the temperature of venous and arterial blood and the lungs. He clearly claimed, however, that venous blood is colder ("this is universally agreed on, and therefore needs no demonstration") an assertion that became part of the "tradition" of animal heat. The instruments were available for direct measurement and a plate picturing one of Fahrenheit's thermometers capable of such is ap-

[16] Boerhaave, *Elements of Chemistry* (1735), pp. 166–167.

[17] It is worth noting that Pitcairne taught at Leyden in 1692–1693 and had several of his books published in Leyden and other Netherlands cities.

[18] For a résumé of Boerhaave's theories of nerve, muscle, and heart action, see E. Bastholm, *The History of Muscle Physiology from the Natural Philosophers to Albrecht von Haller* (Acta Hist. Scient. Nat. Med., vol. 7, Copenhagen; Munksgaard, 1950), pp. 195–198.

pended to the *Elements*.[19] Measurements had been made of the internal body heat as early as 1681 by Borelli.[20] But it might be argued that such measurements were not really necessary for Boerhaave. For if there is truth in his explanation of the production of animal heat through attrition or friction as the blood is passed through the lungs, then obviously the blood would be cooler prior to its passage through the lungs.

This same approach to animal heat was adopted by Stephen Hales. In the second volume of his *Statical Essays*[21] Hales deals at length with the motion of the blood in the body. After showing by experiment that the blood passed through the capillaries of the lungs with greater rapidity than anywhere else in the body, Hales adds that "we may with good reason conclude, that it [the blood] principally acquires its warmth, by the brisk Agitation it there undergoes." He continues, "this we find by daily Experience, that an accelerated Motion of the Blood by Labour or Exercise, does constantly increase its Heat," from which he infers that blood "acquires its Warmth chiefly in the Lungs, where it moves with much greater Rapidity, than in any other capillary Vessels of the Body."[22]

Hales's work, although only published in 1733, was begun some twenty-five years earlier, thus making his ideas contemporaneous with Boerhaave's; and in many ways they are quite similar. Hales also believed that the body heat arose principally from friction and that a strong

[19] *Dr. Boerhaave's Elements of Chymistry, Faithfully Abridged* . . . (London, 1734), Plate 5, Fig. N.

[20] Giovanni Borelli, *De motu animalium* (2nd ed., Lugduni Batavis, 1685), II, 137–138.

[21] Stephen Hales, *Statical Essays: containing Haemastaticks; or an Account of Some Hydraulick and Hydrostatical Experiments Made on the Blood and Blood-Vessels of Animals* . . . (London, 1733), II.

[22] *Ibid.*, p. 90.

brisk motion of the body increases heat much more rapidly than is possible of any effervescent or fermentative motion. And, after all, he pointed out, on cessation of the blood's motion (as in death), the blood cools just as fast as any other equally dense fluid. He suggested that the acquiring of warmth is the principal function of the red globules since they are made to pass through many "converging canals" thus creating great amounts of friction. Hales has not reverted to a wholly mechanical scheme, however, for he suggested that the red color of the globules indicates that they abound in sulphur which makes them more susceptible and retentive of heat. Leeuwenhoek's observation that the blood of fishes, which are cooler than other animals, has a greater proportion of serum in it, and that the blood of a land animal has twenty-five times more globules in it than that of a crab led Hales to conclude that the heat acquired by the body depends either on the different nature or texture of its blood particles or upon the different manner in which the particles act on one another.[23]

A further agreement between Hales's and Boerhaave's theories is in their belief in the dual function of the lungs, refrigerating as well as heating the blood. Hales pointed out that the air is very rapidly heated by the blood in the lungs, for air taken in through the nose will raise the mercury of a thermometer when expired from the mouth.[24]

In keeping with his other attempts to know nature through "number, weight and measure" Hales set about measuring the amount of blood that passed through the lungs in a given period of time. Upon also finding the quantity of air drawn into the lungs he felt that he could

[23] *Ibid.*, pp. 90–92.
[24] *Ibid.*, p. 98.

compute the amount of refrigeration that took place.[25] From this he continued through a lengthy computation of the heat that must, therefore, be added to the blood in its passage through the lungs. He arrived at the figure of 2.98 degrees/minute.[26] Hales quickly added that the blood in the lungs is mixed with the rest of the mass of the blood of the body and thus diminishes the sensible heat passed to the whole body. This last remark plus the one above relating to the heating of the blood indicate that Hales's theory of animal heat suffered in part because he was unable to deal satisfactorily with the accumulation of heat. In the first case there seems to be a confusion between degree of heat and amount of heat, and in the second case he is forced to propose a crude mixing in order to diminish what would otherwise be a destructive amount of heat produced in the lungs.

Hales was already familiar with the experiments that Boerhaave commissioned Fahrenheit to carry out, namely, determining within what degree of heat of the atmosphere an animal could survive. In reinforcing Boerhaave's notion, he concluded that a major use of the lungs must be to refrigerate the blood by inspiring fresh air, for he noted that the natural heat of the blood is not far from the point at which it coagulates, and this would soon be reached were it not for the cooling effects of respiration. He added to this his belief that the observed floridness of arterial blood may be due to the strong agitation and friction undergone in the passage through the lungs. In support of this mechanical coloring of the blood he noted that blood

[25] It is this same question of refrigeration that brought Boerhaave to his discussion of animal heat.

[26] Hales, *Statical Essays*, pp. 98–100. It is not quite clear which scale he is using, but it is probably Fahrenheit's since he has several other references to him.

agitated in a closed glass vessel also became quite florid. He included a typical Halesian remark, probably based on his reading of John Mayow:

'Tis probable also that the Blood may in the Lungs receive some other important Influences from the Air, which is in such great Quantities inspired into them.[27]

He added to his discussion several medical comments, one of which is also quite in character for the country curate. He noted from previous experiments that brandy contracts the fine capillaries of the gut and also thickens the blood and other humors, thus contributing to a sudden increase in the heat of the blood due to increasing the friction in the contracted capillaries.

Hence it is [he said] that the unhappy habitual Drinkers of Brandy and other distilled spirituous Liquours, do so insatiably from time to time thirst to drink of that deadly Liquor, which by often heating the Blood and contracting the Blood-vessels, does by degrees reduce them to such a cold, relaxed and languid State, as most impetuously drives them to seek for their Relief in that Liquor, which they know too well, as by the daily Destruction of Thousands, to be so very baneful and deadly, as to become by the great Abuse of them the most epidemic and destructive Plague that ever befel Mankind.[28]

A similar system of cold contracting and heat dilating he believed was at work in the pores of the body, thus controlling perspiration.

Hales's mechanical theory of production of animal heat suffers no more nor less than the other similar theories proposed at the time. In essence Hales has done little more than state the current (or Newtonian) theory of heat and then plug it into the existing gap in the knowledge of the functioning of the animal body. His solution, like the

[27] *Ibid.*, pp. 104–106.
[28] *Ibid.*, p. 129.

others of its kind, seemed to be precipitated by two over-riding considerations. One was the fact of blood circulation, demonstrated just one hundred years before by William Harvey, which still occupied the research efforts of some of the ablest physiologists, Hales prominent among them. Second was the combined strength of the iatromechanical tradition and the successful Newtonian physics. To mention Boerhaave and Hales is to mention two scientists who consciously attempted to apply the mechanical view to fields far from those in which Newton himself worked.

Like Boerhaave, Hales seems to have consciously turned away from the analogy which had been drawn between combustion and respiration. This is even more striking in Hales's case since we know that he was thoroughly familiar with the writings of John Mayow; indeed, Hales was the one major eighteenth-century student of airs and gases to utilize the findings of Mayow.

Hales noted as a conjecture what to Hooke and Mayow had seemed a certainty, namely, that the blood in its passage through the lungs received some "influences" from the air. But, immediately following this all too brief acknowledgement of the "nitro-aerial particles," Hales concluded that although respiration and its use had been long under investigation science was still in the dark about it. In another closely related "oversight," Hales advocated the view that the floridness of arterial blood was probably owing to the strong agitation and friction that it underwent in the lungs. Here again he overlooked the suggestion of another of the seventeenth-century physiologists, Richard Lower, with whose writings he was quite familiar.[29] Lower, as we have already pointed out, suggested that the color change in blood was due to the addition of the same part

29 *Ibid.*, pp. 105–106, 72.

of the air that others had pointed to as being necessary to combustion.

John Arbuthnot, one of the better-known London practitioners of his day, helped secure the acceptance of a mechanical explanation for animal heat.[30] In his *Essay Concerning the Effects of Air on Human Bodies*, Arbuthnot cites Hales and Boerhaave and then proceeds to adopt their view of the role of the circulatory system in effecting the warmth of animals.[31]

He claims to have found the blood cooler in the veins than in the arteries. The right ventricle of the heart cooled the blood still further, for it was mixed with the chyle. Within the lungs, however, the blood is heated again "so as to render it spumous." Although the cool air does serve to refrigerate the blood, this, Arbuthnot claims, is not the primary function of respiration. The lungs serve instead as the "chief Instrument of Sanguification, working somewhat after the manner of a Press, churning and mixing together the Blood and Chyle."[32]

Arbuthnot mentioned, as did Hales, the then common theory that a mixing of sulphur and air produced fire and was willing to have this action occur in the lungs, but the real explanation for his claim of great heat in the lungs was sought elsewhere. After all, he argued, fishes have more of the salt and oil in their blood than terrestrial animals and yet they are cooler. No, the answer lay in the observation that the blood moves through the small vessels of the

[30] John Arbuthnot (1667–1735), M. D., St. Andrews, 1696; F. R. S., 1704; Fellow Royal College of Physicians, 1710. Physician to Queen Anne and intimate of the literary world of Swift and Pope, Arbuthnot was a popular and witty author as well as a successful practitioner (*DNB*).

[31] John Arbuthnot, *An Essay Concerning the Effects of Air on Human Bodies* (London, 1733, 2nd ed. 1751), pp. 3, 46.

[32] *Ibid.*, pp. 48–49, 98.

lungs forty-three times more rapidly than through capillaries in other parts of the body. His conclusion:

The Heat of the Blood is the Effect of Motion and Attrition of elastick particles, and for that Reason is greater in the Lungs, than in any other Organ.[33]

The fact that the blood contains more saline and sulphureous materials than a mere watery fluid, Arbuthnot added, probably makes it more easily heated through motion.

Another figure who seemed motivated by the combined mechanical legacy to the eighteenth century was George Martine (or Martin).[34] In his first published work he declared his "Assent in general to the Opinion now most commonly received, *That the Heat of Animals is produced by the Motion of the Blood in the Vascular System.*" Martine stated quite clearly that by this he does not mean the "intestine Motion of the Particles of the flowing Blood," but rather that the heat arises in the "Course of the Blood pressing and rubbing upon the Sides of the Vessels." He further claimed that no recourse is necessary to "chemical

---

[33] *Ibid.*, pp. 107–108, 114.

[34] Born in Scotland, 1702; died in America, 1741, while accompanying Lord Cathcart's expedition. Began study of medicine at Edinburgh, 1720, finishing in Leyden with M. D., 1725 (*DNB*). His publications of immediate interest include *De similibus animalibus et animalium calore libri duo* (London, 1740); "Some thoughts concerning the production of Animal Heat, and the Divarications of the Vascular System, being an Abstract from a Latin Treatise of the Heat of Animals; in a Letter to Dr. John Stevenson Physician in Edinburgh," *Medical Essays and Observations* (Edinburgh, 3rd ed., 1747), III, 111–133 (1st ed., 1733–1744); *Essays Medical and Philosophical* (London, 1740), containing "Essay on the Construction and Graduation of Thermometers" and "An Essay towards a Natural and Experimental History of the Various Degrees of Heat in Bodies," both of which were reproduced in a separate volume (Edinburgh, 1772–1787, 1792), with a French edition by Lavirotte (Paris, 1751). Also questioningly attributed to him in the British Museum Catalogue is *An Examination of the Newtonian Argument for the Emptiness of Space and of the Resistance of Subtile Fluids* (1740).

Principles" to support the animal heat; it is not necessary "to suppose Heat to be a sort of animal Process producing a certain Change in the Aliments, some way analogous to the *Luctas* and *Effervescencies* we observe" elsewhere.[35]

The problem which most concerned Martine in this early paper was the similarity in degree of heat in all parts of the body and how it is maintained. That the heat is, indeed, nearly the same in all the parts he claims to "have confirmed by a thousand Experiments."[36] He noted that the heat generally corresponded to the degree of motion of the blood. This, he felt, was certainly aided by the fact that the vascular system was so constructed as to yield a uniform heat everywhere, regardless of the differing speeds of the blood.[37] Although this sounds like a built-in correction, guaranteeing that Martine's system will yield proper results, it is not quite so. Martine is too much of an empiricist. As he was willing to make one thousand measurements of the body temperature, he also made and collected from the literature many measurements of the size and capacity of the arteries.[38] The remainder of his first paper is devoted to a recounting of observations and computations in which the blood is treated like any other liquor, and the blood vessels as canals whose areas must be determined. All of this work, he asserted, rests upon the general supposition

[35] Martine, "Production of Animal Heat," p. 112. See the discussion below of Stevenson, whose ideas are here being criticized.

[36] These measurements, many of which are reported in his *Essays Medical and Philosophical*, served as the basis for much of the discussion of the heat of animals which took place in the decades following their publication. They are among the earliest careful, comparative measurements made of body temperature.

[37] Martine, "Production of Animal Heat," p. 113. The similarity to Pitcairne is marked.

[38] In an interesting table he compares the capacity expected "by the Theory" with the results obtained "by measuring." *Ibid.*, pp. 127, 128.

that the Intensity of Heat generated by Attrition is, *caeteris paribus,* in Proportion to the relative Celerity, wherewith the Bodies rub against one another.[39]

The special case involved is the one in which a liquor is forcibly propelled through a canal.

Martine's works form an interesting combination of careful observation and measurement combined with acceptance of a theory for which he is unable to provide the obvious experiments. While able to give data for all sorts of temperature measurement, Martine nowhere suggests the necessity for measuring the rise in temperature he claims accompanies the attrition and friction generated by the movement of a liquid through vessels. Would he have found a temperature increment? Probably not with the instruments he had available.

Martine's theory escapes from the difficulty that both Boerhaave's and Hales's faced—the lungs as the site of heat production and the consequent intense heat that must in some way be modified. Since it is the liquid itself and not just globules which can cause heat in attrition against the walls of the vessels, and since he found laws which relate the diameter of the different vessels and the velocity of the fluids in them[40] and thus provide for an equalization of heat in the diverse parts of the body, Martine is able to account for the fairly regular temperature of the whole body in a way quite consistent with his general scheme. He avoided the paradox of the superheating and refrigeration of a single organ, the lung. But Martine, like his predecessors, was not able to distinguish successfully between the degree of heat of the animal's body and the

[39] *Ibid.,* p. 113.
[40] *Ibid.,* pp. 118–121.

quantity of heat generated by the motion of the blood. The animal heat he measured was merely the temperature.

Perhaps the most curious attempt to account for the production of animal heat by means of some mechanical friction is that proposed by Robert Douglas.[41] After citing Newton on the role of hypothesis in experimental science, Douglas in his Advertisement proclaimed his "Attempt to explain the Generation of Heat in Animals; not by any preconceived Hypothesis or mere Opinion, but by an easy and natural Deduction from such Premisses, as are solely founded on Observation and Experiment."[42] In his treatment of the material at hand he has "follow'd somewhat of the order of the Mathematicians" which he feels enables him to give more meaning and clarity. The mathematical order consists in first setting out a series of eight "Definitions," following these with "Observations," a report of the "Phaenomena," a group of "Lemmata," a section of "Theorems," and finally a concluding "General Scholium."

Douglas was aware of previous theories of animal heat. He noted that opinion of "the last century" attributed the warmth to an "intestine Motion of the blood," or alternatively to some fermentation, effervescence, ebullition, or accension taking place in the heart or blood.[43] These views were succeeded by the ideas of some sort of mechanical friction and agitation of fluids and solids. This latter he identified as the view of Boerhaave and other European contemporaries. In the text itself the other authors he cites are Newton, Hales, and Martine.

[41] His publications are few but pertinent: a thesis *De incerto indicio velocitatis sanguinis in febribus, ex calore et pulsu,* 6 pp. (Edinburgh, 1746); *An Essay Concerning the Generation of Heat in Animals* (London, 1747), and two editions of the French translation of the same (Paris, 1755, 1760).

[42] Douglas, *Generation of Heat in Animals,* Advertisement.

[43] *Ibid.,* p. 6.

It is worth examining several of the "Definitions" that Douglas provides, for he is desperately trying to bring some order to the language used in discussions of heat of the animal body. He has taken the old, very general term "Innate Heat" and severely limited its meaning. It now refers to the excess of the animal's absolute heat (temperature) over that of the ambient medium. Although he still suffers from the same confusion between quantity of heat and temperature that plagued his contemporaries, he does make an attempt to differentiate between them and in his fourth "Definition" says:

> By the Quantity of Heat which an Animal generates is understood that Supply of Heat continually occupying the Place of the Waste which the Body of a hot Animal necessarily suffers by its immediate Contact with a colder Medium.

This waste, and consequently the amount of heat generated by animals of different sizes, he goes on to explain, "must be compounded in a Ratio directly as their innate Heat, and inversely as their Diameters." In defining the "Intestine Motion of our Fluids" he eschews argument by not restricting it to either fermentation, putrefaction, or effervescence, claiming that "in this Argument, it is none of my Business to determine what particular Modus of an Intestine Motion exists in the Blood." The sole question in which a chemical discussion might have occurred is thus avoided, although later in the text Douglas proved to his own satisfaction that the "Generation of Animal Heat is not the Effect of any intestine Motion in the Blood." This he argued from the definition of animal heat itself, as requiring a certain degree of external cold. "Fermentations," he pointed out, "are retarded by cold, and effervescent Liquours are not influenced by heat or cold."[44]

[44] *Ibid.*, pp. 8, 10, 13, 32, 33.

The "Observations" adduced are of a similar mechanical nature, and in at least one case can hardly be classed as observable. The "friction of the Globules in the extream Capillaries ceases," we are told, "when the Heat of the Animal coincides with that of its medium." This is clearly a deduction from the previous "observation" in which it has been claimed to be known by common experience (and on the testimony of Stephen Hales) that heat relaxes the vessels of the animal. Douglas hastily adds, however, that this is different from the general phenomena of rarefaction and condensation, for the body temperature itself remains the same in hot weather, with the vessels relaxed, as in cold weather, when the vessels are contracted.[45]

Having amassed his "Definitions, Observations and Phae-nomena," the "Proposition" he hoped to demonstrate is "That Animal Heat is generated by the Friction of the Globules of our Blood against the Sides of the extream Capillaries."[46] This can be recognized as a modification of the previous friction theories. The globules forcing their way through the small vessels become the operative factor, but not, as for Boerhaave and Hales, limited in the site of their action to the lungs. Rather Douglas is concerned, as was Martine, with the "extream capillaries" of the whole body.

Douglas is at pains in the four "Lemmata" to disprove several common notions, among them that heat is produced by a mutual friction of the fluids and solids of the blood; that an "intestine motion" was capable of creating animal heat; or that the heat of an animal could be generated by "the mutual Attrition of its Solids on each other" with

[45] *Ibid.*, pp. 19, 16–17, 18.
[46] *Ibid.*, p. 27.

the exception of the action of the globules against the walls of the extreme capillaries. The argument proceeds in a rather traditional manner until Douglas is able to claim, with true mathematical delight, that his Proposition "is an easy Corollary of the four preceding Lemmata." It should be noted that most of the discussion already mentioned, and the further evidence for the "Proposition" put forward by Douglas, relies only in the remotest sense on observation or experiment. His arguments are primarily verbal; the system he constructs is self-correcting; for example, the capillaries relax to a level at which there is no friction of globules when the ambient medium and body are at the same temperature. No experiment or even group of experiments is needed to demonstrate this, since the constriction and relaxation of the vessels is taken as common knowledge according to Douglas. All that need be adduced is that there is no other major source of animal heat, and even this argument proceeds primarily on the verbal level. Douglas, as did Hales and Martine before him, includes some arithmetic among his evidence. In this case he computed the vast surface in which friction can take place, comprising the surface of the globules and the walls of the vessels, and is able to conclude that "the Friction of the Globules in the extream Capillaries not only easily solves the Phaenomena of Animal Heat, but seems a cause fully adequate to the Effect."[47] All this without any independent measurement of how much heat can actually be produced by friction.

The theories proposed in these iatromechanical works found their way into the influential texts and treatises of the mid-eighteenth century. Jean Baptiste Senac utilized

[47] *Ibid.*, pp. 28f, 32f, 36f, 47, 137.

a mixture of theories of this general nature in his volume on the structure of the heart.[48] While admitting that fermentation is the "prime mover" in the body, he claimed that it is the continual movement and agitation of the solids and fluids that produces the body's heat. The observation that as the blood moves more slowly the body heat becomes feebler and that conversely the movement of the arteries always reanimates the heat he saw as a direct demonstration that the heat is produced by the action of the solid parts of the blood.[49]

Senac pointed to the nature of the blood itself, its weight, its density, and its oily parts as the agents through which the arteries are able to produce heat. He cited Martine for his belief that it is the force of the arteries, especially the capillaries of the extremities, as they resist the movement of the heavy blood, which is the main cause of heat. He argued against Willis, who had claimed that no amount of agitation would raise the temperature of a liquid. Similarly, he contradicted Douglas, who had limited heat generation to the interaction of the solid parts of the blood with the walls of the containing vessels. The blood is able to conserve the heat for a longer time than other parts, Senac claimed, owing to the oils in it which "are warmed more than other matter."[50]

In a survey of the degrees of heat maintained by warm-blooded animals, J. A. Braunio provides reports of direct measurements made by inserting his thermometer into wounds in the animal body.[51] Although he concluded that

[48] Jean Baptiste Senac, *Traité de la Structure du Coeur, de son Action, et de ses Maladies* (Paris, 1749, 2nd ed., 1777), vols. I–II.

[49] *Ibid.*, II, pp. 240–241.

[50] *Ibid.*, pp. 243, 244. An echo of Stephen Hales's view of the role of sulphur.

[51] J. A. Braunio, "A review of his *Dissertatio Physica Experimentalis, de Calore Animalium*," *Medical Commentaries*, 1 (1773), 59–62. The

all animals with lungs, including the lungfish, were able to maintain a degree of heat comparable to quadrupeds, Braunio surveyed the current theories of animal heat and chose as most likely the one which attributed heat generation to mechanical friction.[52] No explanation for the choice is offered, and we can only conclude that the mechanical theory of animal heat was not unattractive to the physiologist of the mid-eighteenth century.

When Albrecht von Haller adopted the friction theory of animal heat the idea became embedded in the physiological literature in a manner not to be easily removed. A full one hundred years later James Prescott Joule in a postscript to his paper "On the Calorific Effects of Magneto-Electricity, and on the Mechanical Value of Heat" recorded a conversation with a physician, Dr. Davies, in which he relates that Davies had recently "attempted to account for that part of animal heat which Crawford's theory had left unexplained, by the friction of the blood in veins and arteries." However, Davies did not carry his work far, for he chanced to read a similar hypothesis in Haller's *Physiology*.[53] The text referred to is surely one of the numerous later editions of Haller's *Primae linae Physiologiae* of 1747.

In this famous treatise Haller noted that in warm-blooded animals there is "a most prodigious degree of friction, as well from the blood globules upon the sides of the arteries, as from the arteries themselves contracting round the blood; to which add, the attrition of the particles of blood against each other by the confused and vortical motion

---

Dissertatio originally appeared as part of the *Novi Commentarii Academiae Scientiarum Imperialis Petropolitanae*, XIII.

[52] *Ibid.*, p. 62.
[53] *Phil. Mag.* [3], 23 (1843), 442.

with which they are propelled." He went on to enquire "whether the heat of the blood does not also proceed from its motion? seeing we observe heat to arise from the motion of all kinds of fluids, even of air itself." Haller did not produce any new evidence for his views, but rather proceeded with a common-sense argument. The blood is warm in fish which have a large heart and cold in those which have a small one. He pointed to the more intense heat of birds that have a large heart and rapid pulsations, and to the increase of animal heat that follows from exercise or even from bare friction of the parts. He noted the coldness of people who have a weak and obscure pulse. To make certain that he was not misunderstood, Haller claimed that the heat does not "at all arise at first from any degree of putrefaction in the blood, seeing the humours themselves, when left at rest, generate no heat." And if a chemical action is incapable of producing animal heat, neither "must we explain an evident appearance from the action of such an obscure being as the *vital power*." In an attempt to explain what might otherwise appear as an anomaly Haller added that although "sometimes the heat may be greater when the pulse is slow, and less when it is more frequent, the difference may arise from the different disposition of the blood, from the different densities of the vessels, or the increase or diminution of perspiration."[54]

Haller's system includes several other features of those of his predecessors. Although he believed that the agitation of the blood in the lungs adds greatly to its heat, Haller also recognized that the blood is cooled in the lungs. "But that

---

[54] Albrecht von Haller, *First Lines of Physiology* . . . , translated and printed under the Inspection of William Cullen (Edinburgh, 1786), I, 105–107. Haller has adopted a system almost identical with the earlier one proposed by Pitcairne.

this was not the principal design of nature, is evident," he pointed out, "since no one will say that the venous blood is hotter than the arterial."[55]

By the middle of the eighteenth century there was wide acceptance among physiologists of the belief that animal heat was the product of some sort of agitation, attrition, or friction caused by the blood's passage through portions of the arterial system. This belief seemed to be based more on a reaction to the iatrochemical notions of fermentation, putrefaction, and coction than on any new experimental evidence. This is not to say that interesting experiments were not carried out, or that important measurements of body heat were not made. It would seem that the physiologists mentioned above were more concerned to bring a former "vital" function, animal heat, into accord with the mechanical theories of heat so prominent in the years following Newton.[56] The authority of the Newtonian view and the seeming success achieved by mechanical theories of heat might well be the reason why eighteenth-century physiologists (and chemists) gave up their attempts to explain animal heat by analogy to combustion and shifted instead to what seemed like a simple mechanical function. How convenient it was to treat blood like any other fluid, and the arteries and veins as though they were conduits through which it moved. The early-eighteenth-century mechanical physiologists had been very successful in reducing the blood circulation to a problem of hydrody-

---

[55] *Ibid.*, pp. 157, 160. Haller is here following the assertion of his teacher Boerhaave who claimed that this fact needed no demonstration.

[56] A good case can be made for calling Boerhaave, Hales, Martine, and Douglas eighteenth-century Newtonians. See for example, Chapters 6 and 7 of I. B. Cohen's *Franklin and Newton* (Philadelphia: American Philosophical Society, 1956), for an interesting discussion of the Newtonian impact on eighteenth-century experimental science.

namics. It would seem that they thought they might be equally successful in explaining heat by tying it to the motion of the body's fluids.

The mechanical theories, however, did not pass without criticism. A brief look at this criticism and at the alternatives suggested is important if we are to understand the revolution in biological thought that accompanied the chemical revolution of the late eighteenth century. An examination shows that the most "successful" early-eighteenth-century theories of animal functions were non-chemical ones, an observation which is certainly not true of the latter half of the same century.

In an article of several pages in the famous *Encyclopédie* of Diderot and d'Alembert, Gabriel-François Venel dealt at length with the friction theory of animal heat.[57] While citing Boerhaave and Martine in passing, he turned his attention primarily to the theory as proposed by Douglas. Venel's own definition of animal heat is identical with the definition of "Innate Heat" of Douglas (as reported above) to the extent of using the same numerical example.[58] Venel noted that "les plus célèbres Physiologistes" now explain animal heat as being due to agitation or friction of some sort and he applauded Douglas's criticisms of these various attempts. Venel went on to point out, however, that even the ingenious system of Douglas is nothing more than a

---

[57] Gabriel-François Venel, 1723–1775. Commencing with vol. III, Venel is responsible for most of the articles on physiology, chemistry, medicine and pharmacy in the *Encyclopédie*. While Professor of Medicine at Montpellier he launched a vigorous attack on Boerhaave's method in medicine. See *Éloges des Académiciens de Montpellier*, ed. R.-N. D. Desgenettes (Paris, 1811); also Desgenettes's article in *Biographie Médicale*, VII (Paris, 1825), pp. 407–411.

[58] *Encyclopédie ou Dictionnaire Raisonné des Sciences, des Arts et des Métiers* (Paris, 1753), III, 31; cf. Douglas, *Generation of Heat in Animals*, p. 8.

hypothesis. He further claimed that Douglas's enumeration of the possible causes of animal heat, all of which he has refuted, is incomplete. It has neglected any system which had no recourse to a mechanical cause.[59]

Venel is specifically critical of Douglas's proposal that the vessels relax and constrict in response to changes in the external temperature, and thus provide for just the proper amount of friction. What of the effect of the heat of the body itself on the vessels, asked Venel? He found on reducing Douglas's theory to its simplest statement that the same vessels will at the same time be both relaxed and constricted, both hot and cold.[60]

The telling criticism made by Venel comes in the discussion of Douglas's computation of the surfaces on which friction will take place and the expectations of the heat which will in consequence be evolved. Venel claimed that Douglas confused the quantity of heat with the degree and that these are distinctly separate things. He continued with a lengthy example in which he attempted to show that, as Douglas's globules become heated in passing through the capillaries, they are not able to provide a temperature any greater than that of a single globule, that is, the heat (or really the temperature) of one globule cannot be added to that of another and so on to achieve a very great temperature. Although Venel has provided a much-needed distinction, Douglas does not really seem guilty of the error except by omission. He had avoided the problem simply by not dealing with the question of how much heat

[59] *Encyclopédie*, pp. 33–34. For a discussion of Venel's criticism of physics, see C. C. Gillispie, "The *Encyclopédie* and the Jacobin Philosophy of Science: A Study in Ideas and Consequences," in *Critical Problems in the History of Science*, ed. Marshall Clagett (Madison: University of Wisconsin Press, 1959), pp. 258–260.
[60] *Encyclopédie*, p. 34.

is in reality produced. All he and the other supporters of the friction theory did was to assert that through friction of various sorts the constant temperature of the animal would be maintained. The criticism by Venel is much more applicable to Boerhaave and Hales, who were forced to deal with the great excess of heat which would be produced in the lungs. Venel concluded his article with a general criticism of those who seek mechanical solutions to medical problems.[61]

The alternatives to the mechanical theories of heat offered during the early eighteenth century were for the most part a reversion to earlier chemical ideas with very little new thought or experimentation to back them up. Typical of these is the paper of Cromwell Mortimer included in the *Philosophical Transactions* of 1745.[62] He began by noting the attrition or friction theory of Boerhaave, but went on to claim that no heat is produced by the agitation of fluids "nor can the Blood of Animals, when once let out of the Body, be kept either fluid or warm by [even] the most violent Agitation."[63] However, he

---

[61] *Ibid.*, pp. 34–35. "M. Douglas paroit avoir confondu ici la quantité de *chaleur* avec le degré: mais ce sont deux choses bien différentes . . . Cent globules frottés, ou cent pintes d'eau contiennent une quantité de *chaleur*, comme 100, où sont cent corps chauds; un seul globule, ou une seule pinte, ne sont que la centième partie de cette masse chaude: mais le degré de *chaleur* est le même dans le globule seul et dans les cents globules, ou dans un million de globules. Ainsi si chaque globule ne peut dans son trajet dans un vaisseau capillaire produire sous la température supposée une *chaleur* de 66^d, il est impossible que tel nombre de globules qu'on voudra imaginer produise ce degré de *chaleur*."

[62] "A Letter to Martin Folkes, Esq.; President of the Royal Society, from Cromwell Mortimer, M. D. Secr. of the Same, concerning the natural Heat of Animals," *Phil. Trans. Roy. Soc. (London),* 43, No. 476 (1745), 473–480. Mortimer studied medicine with Boerhaave at Leyden, graduated M. D., 1724, was elected F. R. S., 1728, and served as second or acting Secretary of the Royal Society from 1730 until his death in 1752. He contributed numerous papers to the *Phil. Trans.*

[63] *Ibid.*, p. 474. Mortimer is clearly repeating the argument presented by Willis.

explained, heat is produced in fluids by fermentation and effervescence. Fermentation is defined by Mortimer as a spontaneous intestine motion, confined to the vegetable kingdom. The heat produced by a ferment, he pointed out, never exceeds that of the human body. Effervescence also arises from an intestine motion, either by the mixing together of fluids of different natures or by the addition to fluids of salts or powders of different natures; for example, acids and alkalis on being mixed cause great ebullition, but no heat, while the solution of some metals in *aqua fortis* causes intense heat and emits flame.[64]

Phosphorus was regarded by Mortimer as the "animal sulphur." All animals, he believed, contain some phosphoreal principle; for example, there are insects that shine or emit light, there are luminous fish, and some quadrupeds emit light upon having their hair rubbed. Even the human body can appear luminous. These examples he evoked as proof that phosphorus exists, at least in the dormant state, in the animal fluids. All that is necessary is that the phosphoreal particles be brought into contact with aerial particles and as a consequence heat must be produced. Thanks to the superabundance of aqueous humors in the body, flame is avoided. "This, I think, explains evidently the Cause of animal Heat: Indeed," Mortimer continues, "the Heart and Arteries are the Instruments which excite this Heat; but that is not done by the Friction caused by the Circulation of the Humours, but only by the intestine Motion, which the Circulation gives to the several Particles which constitute the Mass of animal Fluids."[65] As the velocity of the fluids increases, the speed of the particles increases and consequently the phosphoreal and the aerial

[64] *Ibid.*, p. 475. This is a continuation of the ideas discussed in the last chapter and would seem to be specifically influenced by Thomas Willis, who proposed that fermentation involved an intestine motion.
[65] *Ibid.*, p. 477.

particles come into contact oftener, thus generating more heat.

This theory contains some of the elements of the aerial nitre proposals of Hooke and Mayow. It should be noted that Mortimer claims that these ideas reported in 1745 were actually the substance of a letter sent to Boerhaave some twenty years earlier. The wise Master replied only that it was a "pretty hypothesis."[66] Mortimer, it seems, was tempted to revive his theory upon reading in a recent number of the *Philosophical Transactions* of three persons who had died in flames.[67] One of the unfortunates had "charged all her Pores and absorbent Vessels with a great Quantity of Camphire"; the other, it seemed, "had drank plenty of Gin." Both these additions to the animal economy were deemed to have promoted the kindling of the phosphoreal fire within.[68]

The chemical theories offered by John Stevenson in his paper of 1747 are as well reasoned as any proposed during the early eighteenth century.[69] He was familiar with the literature on animal heat and seems to have been the critical reviewer of whom Martine spoke in his own paper in the *Edinburgh Medical Essays*.[70] Stevenson opens his essay with the pessimistic note that the theory of animal heat at the

[66] *Ibid.*, p. 480.

[67] "An Extract by Mr. Paul Rolli, F. R. S. of an Italian Treatise, written by the Reverend Joseph Bianchini, a Prebend in the City of Verona; upon the Death of the Countess Cornelia Zangári & Bandi, of Ceséna. To which are subjoined Accounts of the Death of Jo. Hitchell, who was burned to Death by Lightning; and of Grace Pett at Ipswich, whose Body was consumed to a Coal," *Phil. Trans. Roy. Soc. (London)*, 43, No. 476 (1745), 447–465.

[68] Mortimer, "Natural Heat of Animals," pp. 478–479.

[69] John Stevenson, "Cause of Animal Heat." Stevenson is listed as M. D. He is probably the "Joannes Stevenson, Scoto-Britannus," who matriculated at the University of Leyden, 5 Jan., 1709. He was also a member of the Edinburgh Medical Society.

[70] Martine, "Production of Animal Heat," p. 111.

present time is doubtful at best. One physician he knew believed it was caused by the attrition of solids: another saw an intestine ebullition of the blood as its source. His own attitude is best summed up by his belief that animal functions refuse to be subjected to mechanical laws, "being [instead] formed and conducted by a *divine Set of vital Principles*, an *inward Life and Motion*, which mocks all the bold, vain and frivolous Attempts of our modern philosophy." After listing and subjecting to criticism four already familiar variants of the theory of animal heat through mechanical motion, Stevenson lays the major blame for this outlook on Descartes and his theory that all activities can be reduced to matter and motion. In rather vigorous terms he indicts the mechanical philosophers:

> Not content with the ingenious and useful Application of Levers, Ropes and Pulleys; to the Bones, Muscles and Tendons, and other valuable *mechanical* and *hydrostatical* Pursuits: Not content with these, I say, Millstones were brought into the Stomach, Flint and Steel into the Blood-vessels, Hammer and Vice into the Lungs, &c. But all to no good Purpose; there being certain Bounds beyond which mechanical Principles and Demonstrations do not reach.[71]

Stevenson claims to have measured the temperature of blood himself, collecting it in a cup in which a thermometer was placed. These experiments showed the venous blood to be warmer than the arterial, a direct contradiction of Boerhaave's claim. Perhaps this was more wishful thinking than accurate thermometry (although no less accurate than that of Boerhaave), for it would suit his theory well if the veins were warmer. In the arteries, he points out, the blood meets with nothing which can promote its perfection, "whereas, in the Veins it meets with new Matter . . . par-

[71] Stevenson, "Cause of Animal Heat," pp. 328, 352.

ticularly with new Chyle; and in the Lungs with what we shall not at present give a Name to." As to the attempts to link the florid color of the blood with the idea that it gains its heat in the arteries, he objects: "Indeed, so little convincing has been said on this Head, that one would be almost tempted to imagine, that this Opinion owed its Origin to the Difficulty of separating the two Ideas of *red* and *hot*, which go so oft together."[72] Many of his other criticisms are no less acid and just as pertinent.

Stevenson's own theory is one which he feels has been neglected. He believes that animal heat is evolved through the "animal processes" by which the ailment and fluids are continually undergoing alteration. The bases of his theory are summed up in what he has called several "generally accepted premises:"

1. That the intestine motion of the minute parts of mixed bodies on one another is capable of producing as great or greater heat than the mutual attrition of the surfaces of the hardest bodies;

2. That parts of mixed fluids acting on one another are capable of producing heat;

3. That no attrition of a solid body on a homogenous fluid is known to produce heat;

4. That no alterations of mixed fluids are made by mechanical actions;

5. That the concern with the chemistry of the animal body is as valid as a concern with its mechanics;

6. That most of the chemical processes in which texture is altered are attended by heat.[73]

The animal processes Stevenson has in mind are somewhat analogous to fermentation and putrefaction. Fermen-

[72] *Ibid.*, pp. 332, 342, 341.
[73] *Ibid.*, pp. 353–355.

tation he describes as an intestine motion of the parts of a fluid, not due to either a mechanical action or the influence of the containing vessels. Air is indispensable to the process, either being separated from, or mixing with, the mass. A small but sensible degree of heat is emitted. Putrefaction when carried to its height is accompanied by great heat, smoke, and sometimes flame. Air is necessary, and the texture of the mass is changed.[74]

Similarly, air is necessary for the animal "whether it be in mixing with the Blood, or separating from it, or both." The heat produced, Stevenson claims, lies somewhere between that of fermentation and that of putrefaction, the animal process more nearly resembling the characteristics of the latter. "I don't see more requisite to establish this *animal Process* to be of the same Nature, and indeed to be a certain Degree of the *Process of Putrefaction*." Stevenson offers to the philosopher the very humbling reflection "that *Man's Body, of which he is so vain, is little better than a smoking Dunghill*."[75]

This heat-producing process is located in the animal juices, especially in the blood, "and as there is three or four times more Blood in the Veins than in the Arteries" it seems likely "that this Process is chiefly carried on in the Veins." And to make sure that he will not be misinterpreted, he adds: "and as the specifick Coldness of the Air in the Lungs, seems apter to check than to forward it, it seems probable that there is least of it in the Lungs."[76]

Stevenson is forced to distinguish between fetal heat and animal heat, the former seen as slight and dependent upon the maternal heat. The "animal processes" then have begun

[74] *Ibid.*, pp. 355–356.
[75] *Ibid.*, pp. 356, 357, 362.
[76] *Ibid.*, p. 360.

in an incomplete manner in the fetus and reach perfection only when the animal breathes freely. This should not suggest, however, that respiration is a cause of animal heat for even after "death," when the circulation is stopped, the vital processes continue, though in a less intense manner. An animal, Stevenson claims, "cannot be said to be *dead*, till the Energy of the Blood is so far gone, that though assisted by all possible means, it can never be able again to fill, and stimulate into Contraction the right *Sinus Venosus* and Auricle of the Heart."[77]

Stevenson's criticisms of the mechanical theories of animal heat production are often accurate as well as pungent, but his own theory emerges as a strange mixture of late-seventeenth-century chemistry and vitalism, and, for all its attempts to explain a wide range of animal functions, it remains "merely another hypothesis" with little basis in experiment. But then this lack of experimental basis is a characteristic shared by all the theories so far discussed, for few of them yield to experimental verification. Claude Bernard's suggestion that this was still an era of hypothesis seems fundamentally correct.[78] Even when such consistent experimenters as Steven Hales turned to animal heat, the views they offered fell far short of the demands of eighteenth-century experimental science.

It is not surprising, then, that when other scientists of the early part of the century turned to study the ability of animals to maintain warmth they eschewed the complex mechanical and iatrochemical theories and fell back on a

[77] *Ibid.*, pp. 364–372.

[78] Claude Bernard, *Leçons sur la Chaleur Animale, sur les Effets de la Chaleur et sur la Fièvre* (Paris, 1876), p. 19: "C'est au siècle dernier seulement que la question est entrée dans la voie de l'expérimentation et a quitté la voie périlleuse des hypothèses."

simple vitalism. John Hunter, the surgeon and anatomist, concluded after a series of experiments that

> this power of generating heat seems to be peculiar to animals and vegetables while alive. It is in both a power only of opposition and resistance; for it is not found to exert itself spontaneously and unprovoked; but must always be excited by the energy of some external frigorific agent. In animals it does not depend on the motion of the blood, as some have supposed, because it belongs to animals who have no circulation; besides, the nose of a dog, which is nearly always of the same heat in all temperatures of the air, is well supplied with blood: nor can it be said to depend upon the nervous system, for it is found in animals that have neither brain or nerves.

Hunter believed that even plants and cold-blooded animals maintained a temperature slightly in excess of the environment. They were unable, however, to retain a uniform degree of heat. The power to generate heat was viewed as being in proportion to the "perfection" of the plant or animal. And further,

> it is . . . most probable, that it depends on some . . . principle peculiar to both, and which is one of the properties of life; which can, and does, act independently of circulation, sensation, and volition; *viz.* that power which preserves and regulates the internal machine, and which appears to be common to animals and vegetables.[79]

While Hunter's experiments were of great interest in showing the effects of cold on a wide range of plants and animals, and also the ability of cold to produce internal body changes,[80] his interpretation of his results added noth-

[79] John Hunter, "Experiments on Animals and Vegetables, with Respect to the Power of Producing Heat," *Phil. Trans. Roy. Soc.* (*London*), 45 (1775), 457, 452–454.

[80] See his second memoir, John Hunter, "Of the Heat, &c of Animals and Vegetables," *Phil. Trans. Roy. Soc.* (*London*), 48 (1778), 7–49.

ing new and left the discussion of animal heat in as confused a state as before.

The years following the mid-century saw a proliferation of attempts to provide an explanation for animal heat. Many of the ideas put forward were hybrids of the major proposals, while others attempted to add still new elements to the discussion. The polymath Benjamin Franklin was troubled by the problem of how animals maintain their internal warmth, since the atmosphere about the animal is cooler than the animal itself, and therefore the animal is continually being cooled. Yet the animal's temperature is not lowered, which implies that it is constantly acquiring heat. In a series of experiments Franklin related that he had been unable to produce heat by the friction of fluids, which forced him to conclude that animal heat must be accounted for in some other manner.[81] Indeed, Franklin muses, "how a living animal obtains its quantity of this fluid called fire, is a curious question . . . and I have sometimes suspected that a living body had some power of attracting out of the air, or other bodies, the heat it wanted."[82] What is of great interest in this statement is Franklin's treatment of the heat of animals as a fluid fire. The idea of the fluid fire almost certainly came from Boerhaave—but not from Boerhaave's discussion of animal heat!

Franklin is inclined to the view that fluid fire as well as air is attracted by plants in their growth and is consolidated

[81] I. B. Cohen, *Franklin and Newton*, p. 332. Also Benjamin Franklin, *Experiments and Observations on Electricity, Made at Philadelphia in America* (London, 1769), p. 449.

[82] A Letter to Dr. John Lining in *Benjamin Franklin's Experiments; A New Edition of Franklin's 'Experiments and Observations on Electricity'*, ed. I. B. Cohen (Cambridge, Mass.: Harvard University Press, 1941), p. 342.

with the other materials that make up the plant substance.[83] In turn the plant is digested by the animal "in a kind of fermentation" in which part of the fire as well as the air recovers its active, fluid state. The fire thus reproduced by digestion and separation is continually leaving the body, its place being taken by fresh quantities which arise from the constant separation. Furthermore, Franklin points out, "whatever quickens the motion of the fluids in the animal [for example, exercise], quickens the separation, and reproduces more of the fire."[84]

Thus, Franklin concludes (drawing an analogy to a well-known chemical process):

I imagine that animal heat arises by or from a kind of fermentation in the juices of the body, in the same manner as heat arises in the liquors preparing for distillation, wherein there is a separation of the spirituous, from the watery and earthy parts.—And it is remarkable, that the liquor in a distiller's vat, when in its highest and best state of fermentation, as I have been informed, has nearly the same degree of heat with the human body; that is, about 94 or 96.[85]

To each his own analogy, Stevenson the dunghill, Franklin the fermenting liquid. The new element that Franklin brought to the discussion of animal heat is the idea of a heat fluid. How much easier it is to talk of quantity of heat when a fluid is visualized. However, the quantity could still not be measured. Franklin's theory of fermentation

[83] *Ibid.*, p. 343. This sounds very similar to Stephen Hales's theory of plant nutrition in the *Vegetable Staticks* (London, 1727), p. 320. Hales, however, who thought of heat as motion, would not have conceived of fire as being attracted by plants. Cohen has demonstrated that Franklin was reading Hales in the 1740's. I. B. Cohen, *Franklin and Newton*, p. 276.
[84] Franklin, "Letter to John Lining," p. 343.
[85] *Ibid.*

added nothing new; the idea of a fluid fire being attracted along with the air is of no use until the fire is transferred from its position as a plant nutrient to an element in animal respiration, and this Franklin has not done. However, his idea has demonstrated the versatility of a fluid heat.

Others proposed theories with even less to recommend them. John Caverhill, the Scottish physician, writing in 1769 speculated that the material of the nerves was earthy, and that as this earthy material descended through the nerves to form the bones it raised a certain degree of friction with the sides of the nerve tubes, thus producing the animal heat.[86] In 1770 Caverhill published a second volume "in which," claimed his tender-hearted biographer in the *Dictionary of National Biography*, "he attempted to prove his theory by a large number of barbarous experiments on rabbits, destroying various nerves or portions of the spinal cord, and awaiting the death of the animals." His own explanation of his activities was a little less prosaic. In an effort to determine if his hypothesis was correct Caverhill reasoned that "if nerves were really the cause of heat in animals, every animal deprived of the influence of a certain number of nerves" would at the same time be deprived of that amount of heat which they would normally have contributed to the general heat of the animal.[87] The pattern of experiments is obvious, as will be the outcome, to anyone today with physiological training. What Caverhill succeeded in doing, however, was to introduce one more variable into the already complex question of the generation of animal heat. Contemporaneously with Caverhill, and in

[86] John Caverhill, *A Treatise on the Cause and Cure of the Gout* (London, 1769), p. 25 and *passim*.

[87] John Caverhill, *Experiments on the Cause of Heat in Living Animals and Velocity of the Nervous Fluid* (London, 1770), p. 2.

the years following, there developed a whole school who sought to reserve a place for the nerves as the creators of animal warmth.[88] It is not surprising that physiologists turned to the nerves, for the nerves permitted the type of simple, direct experimentation upon the animal that was denied to those who proposed mechanical or chemical hypotheses. However, the nervous system, which was so closely identified with animal life, could never be accepted as the source of animal heat by the physiologist who was attuned to the natural philosophy of the eighteenth century.

To the observer distant in time (and to some contemporaries) there seems to be little but chaos among the theories of animal heat in the first half of the eighteenth century. Claude Bernard claimed that the lack of proper thermometric instruments in the hands of the physiologists was responsible for the late arrival of the experimental study of animal heat. Physiology, he said, had to await the progress of physics in general.[89] But that progress did not come all at once. During the early part of the century the physiologist had instruments to measure temperature, or degree of heat, and made excellent use of them. Most of the time, however, it was not temperature alone in which he was interested, but rather the total amount of heat the animal produced, allowing it to maintain a constant temperature. Thus the physiologist of the early eighteenth century continually confused thermal intensity with ther-

[88] *Encyclopaedia Britannica* (1st ed., Edinburgh, 1771), I, 32–33. See article on "Aether" for a lively review of an attempt to link the nerves to the production of animal heat. The editors conclude that, in regard to the author of the treatise, "It is, perhaps, wrong to say that he was *reasoned;* for the whole hypothetical part of his essay is a mere farrago of vague assertions, non-entities, illogical conclusion, and extravagant fancies."

[89] Bernard, *Chaleur Animale*, p. 20.

mal content or capacity. Physics had as yet provided neither the theory nor the instrument which could successfully distinguish the two.

It is against this background of confusion in the physical theory of heat, and an equally ill-defined chemistry of fire and combustion, that the development of a chemical theory of animal heat must be viewed. During the first seventy-five years of the eighteenth century the physiologists may have chosen the rather crude analogy of heat from friction to account for the warmth of the animal, but following the advances in physics of heat and chemistry of gases made during the course of the century that analogy was completely (or almost completely) ignored during the last quarter of the same century.

Recognition should clearly be made, however, of the dichotomy in the explanation of biological phenomena that became apparent in the first half of the eighteenth century. The supporters of the friction or mechanical-agitation theory of animal heat had made the fundamental assumption that the living organism did not need special theories and concepts to explain its activities. Biological laws and explanations in their view need not differ substantially from the laws and explanations of the physical world. Their theory, in this case, was poor, their mechanical explanation crude, and in the long run their ideas seem to have contributed little to the understanding of the process under study. Their critics, however, make them important by the sharpness of the distinction indicated as to what is acceptable as an explanation in biology. Stevenson put the point bluntly: animal functions cannot be subjected to mechanical laws, they are instead developed upon a "divine set of vital principles." Analogies to the physical sciences, he would claim, are tenuous at best and the search for understanding should in reality be a search for special biological laws and explanations. It is this assumption which had been unsuccessfully challenged by the iatro-physical and mechanical theories of the late seventeenth and early eighteenth centuries but which ultimately gave way to the theories that emerged from the chemical revolution of the last decades of the eighteenth century.

· V ·

Respiration and Combustion

The two new features which became an essential part of the theory of animal heat during the last two decades of the eighteenth century were the ability to relate heat intensity to heat capacity and the ability to demonstrate experimentally the similarity between respiration and combustion. The importance of Joseph Black's contributions to both of these problems has been shown by several authors.[1] What is of particular interest to us is the form in which these ideas were taken up and developed in the attempt to solve a physiological problem.

It is almost impossible to determine just what Joseph Black's ideas of animal heat were like. His students gave somewhat conflicting reports, and Black's own posthumously published *Chemical Lectures*[2] refers the reader to Adair Crawford's "treatise on this subject, which contains experiments made with amazing labour and much ingenuity."[3] But we are sure of the fact that Black did

[1] Douglas McKie and N. H. de V. Heathcote, *The Discovery of Specific and Latent Heats* (London, 1935); J. R. Partington and Douglas McKie, "Historical Studies on the Phlogiston Theory.—III. Light and Heat in Combustion," *Annals of Science*, 3 (1938), 337–371; Henry Guerlac, "Joseph Black and Fixed Air. A Bicentenary Retrospective, with Some New or Little known Material," *Isis*, 48 (1957), 124–151, 433–456.

[2] Joseph Black, *Lectures on the Elements of Chemistry*, ed. by John Robison (Edinburgh, 1803), vols. I–II.

[3] *Ibid.*, II, 206; see also I, 82. Crawford's work is discussed in detail below.

advance a theory of his own from the evidence in his students' work as well as from the writings of those who in proposing a new theory attempted to refute Black's efforts.[4] One student, Patrick Dugud Leslie, described Dr. Black's contribution as "perhaps the most ingenious and best supported theory which has ever been proposed on the subject of animal heat." Leslie further noted that Dr. Black's view "though never published, is well known to all who have attended his lectures. Besides, I had once the satisfaction of a private conversation with the Doctor on the subject."[5]

Daniel Rutherford, in his "Inaugural Dissertation," provided one of the earliest discussions of animal heat which had been directly influenced by Black.[6] In this interesting paper Rutherford seems to place both fixed air and nitrogen under the head of "mephitic air." He believed that the latter of the two although arising from the lungs of animals is not generated there, but more probably is formed from the food in the body, and is ejected through the lungs as something noxious. He further observed

that the warmer animals are, the more perfect and constant is their respiration, and the more quickly do they infect the air with a malignant nature; may we not then suspect that animal heat and that alteration of the air arise from the same cause?

It is at this point that "Prof. Black" is cited in a footnote as the source of the above conjecture. Rutherford continued:

[4] See, for example, P. Dugud Leslie, *A Philosophical Inquiry into the Cause of Animal Heat* . . . (Edinburgh, 1778), pp. 75ff.

[5] *Ibid.*, pp. 75, 89.

[6] Daniel Rutherford, *Dissertatio Inauguralis de Aere Fixo Dicto, aut Mephitico* (Edinburgh, 1772). A translation by Crum Brown was communicated as "Daniel Rutherford's Inaugural Dissertation," *J. Chem. Educ.*, 12 (1935), 370–375. An even earlier dissertation credited as being a source for Black's theory of animal heat is that of James Maclurg, *Tentamen Medicum Inaugurale, de Calore* . . . (Edinburgh, 1770). Maclurg, a Virginian who had studied in Edinburgh with William Cullen and Joseph Black, received his M. D. from Edinburgh in 1770. His work has been cited by both P. D. Leslie and H. Moises.

As the life of animals depends on the free use of air, so this is altogether necessary for the support of flame and fire. But no less by fire than by respiration is it so changed as to be unfit for either use and contrary to both. And as the effects are nearly the same, what I have brought forward as to respiration may be repeated as to combustion.[7]

The concept is not startlingly new. The ancients had recognized that where a fire could not burn an animal could not live. Robert Boyle, Stephen Hales, and others had realized that a similar alteration of the air (in terms of its volume or "elasticity") was caused by combustion and respiration.[8] It remained for Joseph Black to provide the experimental means of recognizing the similarity of the products of fire and breathing, the gas which he described as "fixed air." But only a partial characterization of "fixed air" and his experimentation with it were published by Black in 1756.[9] His reflections on animal heat and its connection with the intensity of respiration come from the period just following this publication and were never put in print by Black.[10]

It was at this time, 1756 or 1757, that the important researches on heat were begun. From these emerged the doctrine of latent heats of change of state and then the realization that different substances have characteristically different capacities for heat.[11] These ideas, and any further

[7] Rutherford, "Inaugural Dissertation," pp. 373–374.

[8] It is interesting to note that the author of the anonymous *Enquiry into the General Effects of Heat; with Observations on the Theories of Mixture* (London, 1770), an early report of Black's theory of heat, found it "surprizing [that] an air, frequently breathed, and thence rendered unfit for respiration, is likewise totally unfit for inflammation," p. 79.

[9] Joseph Black, "Experiments upon Magnesia Alba, Quicklime, and Some other Alcaline Substances," *Essays and Observations, Physical and Literary. Read Before a Society in Edinburgh*, II (1756), pp. 157–225.

[10] Guerlac, "Joseph Black," pp. 452f.

[11] Letter, Black to Watt, Edinburgh, 15th March, 1780. "I began to give the doctrine of latent heat in my lectures at Glasgow in the winter 1757–58, which I believe was the first winter of my lecturing

speculations Black may have made about animal heat, became a part of the lectures he gave first at Glasgow and then at Edinburgh.[12] The posthumous collation of these lectures indicates many instances of Black's concern to apply the findings of his study of heat to the understanding of biological phenomena. He found heat "inseparably necessary to the very existence of vegetables and animals." He utilized his theory of latent heat in an attempt to explain how animals endure high atmospheric temperatures. He believed "that the heat absorbed in spontaneous evaporation greatly contributes to enable animals to bear the heat of the tropical climates."[13]

Of equal importance with the theoretical contributions to heat studies was Black's development of the technique of comparing relative heat capacities. The method of mixtures was not his invention, having been used earlier by Renaldi in the late seventeenth and Fahrenheit in the early eighteenth century, but its evolution into a calorimetric

---

there; or, if I did not give it that winter, I certainly gave it in 1758–59, and I have delivered it every year since that time in my winter lectures . . . at Glasgow until winter 1766–67, when I began to lecture in Edinburgh." James Patrick Muirhead, *The Origin and Progress of the Mechanical Inventions of James Watt, Illustrated by His Correspondence with His Friends and the Specifications of His Patents* (London, 1854), II, 119.

[12] Some indication of the reception of these ideas by the students who listened may be gained by a glance at the number of dissertations from this period on heat and animal heat which had the Scottish universities as their place of origin. In addition to those noted above a partial list includes: G. R. Brown, *De ortu animalium caloris* (Edinburgh, 1768); P. Dugud [Leslie], *De caloris animalium cause* (Edinburgh, 1775); J. Garner, *De calore animalium* (Glasgow, 1755); C. A. Graham, *De calorico et de evolutione ejus in animalibus* (Edinburgh, 1793); R. M. Hawley, *De fonte caloris animalium pulmonibus instructorum* (Edinburgh, 1807); J. Luby, *De ortu caloris in animalibus respirantibus* (Edinburgh, 1803); G. S. Mitchell, *De animantium calore* (Edinburgh, 1800); G. C. De La Rive, *De calore animali* (Edinburgh, 1797); G. Saunders, *De natura calorici; et de calore animalium* (Edinburgh, 1802).

[13] Joseph Black, *Lectures*, I, 243, 214. This is the problem that had originally been raised by Boerhaave. See Chapter IV above.

device was certainly due to Black's inspiration. From about 1763 forward Black's lectures contained not only the discussion of "fixed air," but also his observations on heat including the method of mixtures.[14]

But Joseph Black's was not the only influence felt by the student at the Scottish universities. If the work of William Irvine on latent heat seemed in general concord with that of Black,[15] the proposals of William Cullen and Andrew Duncan were of quite a different nature.

Cullen, with whom Black had studied at Glasgow, rejected the currently accepted notion that the heat of animals was due to motion or attrition of the parts of the blood. But he was forced to comment that "although it be easier to show, that animal heat is not produced properly from the friction of the fluids upon the vessels, than to substitute a more satisfactory and probable explication, my own reading and experiments, nevertheless, will not allow me to favour this theory." He proposed instead a complex theory involving respiration, the quality and accessibility of the air, the motion of the muscles, and the "progressive" motion of the blood, all directed by "various states and quantity of animal power in the nervous system."[16]

[14] See Henry Lord Brougham, *Lives of Men of Letters and Science, Who Flourished in the Time of George III* (London, 1845), p. 340.

[15] This is testified to by Adair Crawford, *Experiments and Observations on Animal Heat, and the Inflammation of Combustible Bodies. Being an Attempt to Resolve these Phaenomena into a General Law of Nature* (London, 1779), p. 17. Irvine was a student of Black's and also collaborated with him on a number of experiments.

[16] This appears as a footnote to Cullen's edition of Albrecht von Haller's *First Lines of Physiology* (2nd ed.; Edinburgh, 1786), p. 106, n. Charles Blagden reports Cullen's belief that life had a power of generating heat that was independent of any common chemical or mechanical means. Cullen was also credited with the view that animals possessed the power of "generating cold," when atmospheric heat exceeded the normal body temperature. "Experiments and Observations in an heated Room," *Phil. Trans. Roy. Soc. (London)*, 65 (1775), 112.

Putrefaction and "chemical" mixture were carefully ruled out as causes of animal heat as finding no support in either observation or analogy. The motion of the blood, however, was regarded as important, for "we perceive the heat to be increased or diminished, as various causes increase or diminish the motion of the blood." The generating power was not, however, to be found in any small portion of the animal system (such as the lungs), but was thought to be diffused in some unequal manner through the body.[17] The motion theory was not without its difficulties with the varied sizes, ages, and temperaments of animals, and it was for this reason that Cullen was forced to fall back upon some "vital principle of animals," acting through the nervous system, to coordinate all the factors and account for the similarity in heat of otherwise dissimilar animals.

Andrew Duncan was invited to teach the "Institutions of Medicine" at the University of Edinburgh during the winter of 1774.[18] In the course of the lectures he presented, it is reported (seemingly by Duncan himself) that he endeavored to refute the commonly held theories of animal heat, that is, mixture, putrefaction, friction, and so forth. It is also noted that he "refuted" the doctrines taught by "those ingenious and learned professors, Dr. Cullen and

[17] William Cullen, *Institutions of Medicine, Part I, Physiology* (Edinburgh, 1772), pp. 218–219, 220, 222–223.

[18] Andrew Duncan was born at St. Andrews, 1744, matriculated at St. Andrews University, 1759–60, graduated M. A., 1762. He proceeded to Edinburgh for medical studies and in 1764 was elected president of the Medical Society at Edinburgh. In 1769 he received the M. D. from St. Andrews and in 1770 was a licentiate of the Royal College of Physicians in Edinburgh. In 1773 he founded and edited the periodical *Medical and Philosophical Commentaries.* After several frustrated attempts he was elected Professor of the Institutions of Medicine at Edinburgh in 1790. For fourteen years previous to this he had given extramural lectures on the same subject. He died in 1828. See Robert Chambers's *Biographical Dictionary of Eminent Scotsmen,* Vol. I (Glasgow, 1840).

Black; the first of whom imagined that animal heat was to be explained from oscillations excited in the nervous fluid, while the last referred it to the effect of respiration on the living system."[19]

Duncan's own hypothesis attributed animal heat to the evolution of phlogiston, or the principle of inflammability, contained in the blood. This process, which he believed to be brought about by the action of the blood vessels, rested on several points which he reported he had tried to demonstrate. First, the blood contains phlogiston which is evolved, extricated, or brought to a state of activity and motion by the action of the blood vessels to which it is subjected during the circulation. Further, "the evolution of phlogiston is a cause which, through nature, produces heat, whether that heat be apparently excited by mixture, fermentation, percussion, friction, [or] inflammation." Lastly, the heat thus produced is equal to the highest degree of heat which animals in any case possess.[20]

This theory, as proposed by Duncan, appears to be the earliest attempt to account for animal heat through the agency of phlogiston.[21] It will be recalled that Georg Stahl and other early phlogistonists did not give a "chemical" explanation of animal heat and respiration. Duncan records that among those attending his lectures was Patrick Dugud Leslie, who accepted his theory and made it the subject of his doctoral dissertation.[22] Leslie in his treatise of 1778 confirmed this influence and said of Duncan's theory:

[19] Included among a group of "Medical Notes" in Duncan's journal, *Medical and Philosophical Commentaries*, 6 (1779), 99.

[20] *Ibid.*, pp. 99, 100.

[21] A thorough check of the writings of mid-eighteenth-century physicians might reveal precursors.

[22] Patricus Dugud, *De caloris Animalium causa* (Edinburgh, 1775); see also [Duncan], *Medical and Philosophical Commentaries*, 6 (1779), 101–102.

This opinion was first, I believe, explicitly delivered in the university of Edinburgh, during the winter of 1774–5, by my learned and ingenious friend Dr. *Duncan,* whose lectures on the Institutions of Medicine I had, at that time, the pleasure of hearing.

In accepting it as his own, Leslie referred to Duncan's hypothesis as one "which to me seems incumbered with fewer objections, more conformable to the simplicity of nature, and more consentaneous to sound philosophy, than any, which it has been my fortune to hear." But, as though not to heap too much praise on Duncan alone, Leslie noted that theories similar or almost similar had been proposed by Benjamin Franklin and Cromwell Mortimer.[23]

The closeness of the schemes of Leslie and Duncan is evidenced by the five points which Leslie felt he must demonstrate as the bases of his own theory of animal heat: (1) blood contains phlogiston; (2) the phlogiston is evolved by the action of the blood vessels; (3) the evolution of phlogiston is attended by heat; (4) the heat thus generated is sufficient to account for the heat of living animals; (5) the various phenomena of animal heat give proof to all the points enumerated. These are identical to the points which Duncan claimed as his own. Despite his obvious debt to Duncan, Leslie claimed that, of all the theories proposed, Black's was nearest to his own, but with one fundamental difference. Through criticism of Black's views Leslie believed he had found "one of the best means of developing some of the radical principles of [his] present inquiry."[24]

Leslie's volume of 1778 represents in substance the views

[23] Leslie, *Philosophical Inquiry,* pp. 92–94; see Chapter IV above. Leslie was a student at the University of Edinburgh, 1773–1775, M. D., 1775, was elected F. R. S. in 1781, and practiced as a physician in Durham.

[24] *Ibid.,* pp. 100–101, 98–99, 90.

published in his inaugural dissertation in 1775.[25] A third of the book is devoted to spirited criticism of the theories of animal heat proposed by his predecessors and contemporaries, for which Leslie urged the reader's indulgence, claiming that this was done to show why they have been rejected in favor of his own system.[26] Leslie was particularly annoyed with those physicians who "became intoxicated with the conceit, that all the phaenomena of the animal economy were explicable on mechanical principles." The animal body, he claimed, "admits not the solution of all its phaenomena on any particular set of principles." Certainly there are chemical and mechanical causes "but it is not less certain, that many functions peculiar to life are regulated by laws steady and uniform in their operation, which are not dependent on any principles of mechanics, or chymistry, hitherto acknowledged."[27] Although he left room for the influence of a "vital power," Leslie had little recourse to it. His system leans heavily on a chemical analogy.

The difference between Leslie's and Black's views of animal heat is fundamental. While Black believed that respiration was necessary for the generation of animal heat, Leslie saw it as the chief means of cooling the heat of the body. Leslie claimed that they both believed in a principle of inflammability but that Black limited the generation of heat to the lungs where air acted on phlogiston (the inflammable principle) in a manner similar to actual in-

[25] P. Dugud, *Caloris Animalium*. His dissertation was published under his Christian names, Patricus Dugud.

[26] Leslie, *Philosophical Inquiry*, p. 8. We will not deal here with these comments.

[27] *Ibid.*, p. 6. This is a recurrent theme in the history of biology. Physics and chemistry may be the tools of examination, the biologists might claim, but the organism must surely be governed by laws of its own.

flammation, whereas Leslie believed that the animal heat was a "necessary consequence of the constant progressive mutations of the mass of blood" causing the evolution of phlogiston throughout the blood system.[28] Leslie cited Black as specifically believing that a mephitic, phlogisticated air is generated both in the process of respiration and in the burning of a fuel, this alteration of the air being due to its combination with phlogiston. Leslie assented to this[29] but went on to doubt that the lungs are the source of the heat, citing as evidence the belief that even animals without respiratory organs generate some little heat.[30] He sought further support in the work of Stevenson and Cullen, who claimed that the venous blood is warmer than the arterial, which situation could not occur if the blood received its heat in the lungs.[31]

In another argument, from "common observation," Leslie suggested that if the blood were heated in the lungs it would need less of their function in warm weather. Experience, he claimed, shows that in just such circumstances animals breathe more fully and more rapidly in order to cool themselves. Leslie concluded this line of argument with the suggestion that it was more reasonable to assume that respiration carries off some of the phlogiston which has been evolved in the blood, thus serving to

[28] *Ibid.*, pp. 98–99; see also p. 75.

[29] *Ibid.*, p. 76.

[30] *Ibid.*, p. 78. This is probably a reference to the findings of John Hunter, "Experiments on Animals and Vegetables, with Respect to the Power of Producing Heat," *Phil. Trans. Roy. Soc. (London)*, 45 (1775), 447. The problem of the site of heat production was a troubling one up to modern times. See Everett Mendelsohn, "The Controversy Over the Site of Heat Production in the Body," *Proc. Amer. Phil. Soc.*, 105 (1961), 412–420.

[31] Leslie, *Philosophical Inquiry*, p. 79. See Stevenson and Cullen above.

diminish the heat of respiring animals. He cited the analogy used by Duncan, who had compared respiration to the blowing of bellows on a hot body.[32]

Another claim attributed to Black for the first time is that the fetus generates no heat of its own but relies solely on maternal heat. Leslie took issue with this contention and produced several negative arguments, the most important of which is the evidence that late duck embryos produce their own heat for several hours after removal of the egg from the hen even though the lungs are not yet active. "This fact," he claimed, "at once overturns Dr. *Black's* hypothesis, since it affords the clearest evidence of heat being generated in the animal machine before the lungs come into play." Thus, he added, "tho' breathing animals are the warmest, there is not so much reason for saying that they are so, because they breathe, as that they breathe because they are warmest."[33]

Leslie cited as conclusive arguments for the opinion that phlogiston is released from the blood (and is contained in perspiration), first, the fact "that if a piece of silver be kept in contact with our skin, it will be as soon blackened, as if it had been exposed to the principle of inflammability escaping from bodies in any phlogistic process"; and secondly, Dr. Priestley's finding "on putting pieces of the crassamentum of sheep's blood into dephlogisticated air, that in the space of twenty-four hours, so much phlogiston was communicated to this purest of all kinds of air, that it

[32] *Ibid.*, pp. 80–81. Compare this with the evidence provided by Adair Crawford, "Experiments on the Power that Animals, when placed in certain Circumstances, possess of producing Cold," *Phil. Trans. Roy. Soc. (London)*, 71 (1781), 479–491.

[33] Leslie, *Philosophical Inquiry*, pp. 81–82, 86–87. Stevenson, "Cause of Animal Heat," is again cited.

became unfit for respiration." To this Leslie added his own belief that the phlogiston enters the body through the nutriment, from which it is extracted by the digestive power and thence taken up by the lacteals and carried into the circulation where he claimed it to be plentiful.[34]

In the course of his attempt to show that the blood vessels evolve phlogiston in a separable and active state, Leslie entered into a review of theories of the color of the blood and of the respiratory function. He showed a familiarity with the works of Cigna, Hewson, Lavoisier, and Priestley among his contemporaries, and expressed dissatisfaction with the explanations they provided.[35] He noted the recent findings that following both combustion and respiration the atmospheric air is converted into "fixed" or "mephitic" air. Although a flame goes out and respiration ceases when the surrounding medium has become saturated with phlogiston, the death of an animal, he claimed, is not due to the lack of some *pabulum vitae,* but rather occurs because the air is overcharged with phlogiston "which we have reason to believe acts immediately on the nervous system; as all animals that do not presently expire in phlogisticated air, die in violent convulsions." Leslie remained satisfied with the belief that the "sole and ultimate purpose of respiration is to carry off from the body the phlogiston, which the circulat-

[34] Leslie, *Philosophical Inquiry,* pp. 127–128. Joseph Priestley, "Observations on Respiration, and the Use of the Blood," *Phil. Trans. Roy. Soc. (London),* 66 (1776), 226–248.

[35] Leslie, *Philosophical Inquiry,* pp. 129–145. Giovanni Cigna, "De colore Sanguinis experimenta nonnulla," *Misc. Taurin.,* 1 (1759), 68–74; William Hewson, *An Experimental Inquiry into the Properties of the Blood. With Remarks on Some of its Morbid Appearances . . .* (London, 1771); Joseph Priestley, *Experiments and Observations on Different Kinds of Air* (London, 1775–1777), vols. I–III; although he mentions Lavoisier's name numerous times, Priestley does not give any citation.

ing powers are perpetually evolving from the general mass of blood, and reducing to an active state."[36]

The color change that the blood undergoes was seen as further evidence of the above use of respiration. The greatest change in color, Leslie said, occurs when the blood passes from the arterial to the venous system in the extremities. This is brought about by the action of the circulation in evolving the principle of inflammability.[37] It is interesting to note how a strongly held hypothesis can cause the same evidence to be treated in quite different ways. The blood obviously changes color twice; Leslie the phlogistonist believed that the systemic change was the important one. The adherents of both the respiration and mechanical friction theories proposed to look to the alteration undergone in the respiratory organs for the significant physiological functions.

Leslie was proud to note that Priestley, in a paper on respiration, suggested, as he had done somewhat earlier in his *Dissertation* of 1775, that respiration is a phlogistic process. He cites Priestley's conclusion that

one great use of the blood must be to discharge the phlogiston, with which the animal system abounds, imbibing it in the course of its circulation, and imparting it to the air.

However, Leslie took the strongest exception to Priestley's view that phlogiston is "imbibed" by the blood. He wondered how a man of Priestley's "sagacity" could have so reversed the system. For it is through the blood alone, claimed Leslie, that the animal economy can be supplied with phlogiston, which has come into the body with the nutriment and is thence passed to the blood for distribu-

[36] Leslie, *Philosophical Inquiry*, pp. 146, 149–150, 157. In support Leslie cites Fontana, who sees phlogiston destroying life in the same manner as an electric shock.

[37] *Ibid.*, p. 158.

tion. Phlogiston, Leslie concluded, is the nutritious matter to which the animal machine owes its "accretion, vigour, and constant support."[38]

Leslie demonstrated to his own satisfaction that the evolution of phlogiston is attended by heat, through utilizing a series of assertions, arguments, and citations. Bodies which contain most of their phlogiston in a loose and separable state, he claimed, are easiest inflamed and soonest consumed, and emit a great deal of light. Query eighteen of Newton's *Opticks* is cited in an effort to show that Newton relied on the aetherial medium to account for the production, duration, and propagation of heat. Furthermore, Leslie continued,

> since it appears from numberless facts and phaenomena, that his aether is the same principle as the phlogiston of the chymists, it by consequence follows, that the tenets of the chymical and mechanical philosophers, on the subject of heat, are by no means so widely different.

Heat, he found, is ultimately excited by the evolution or extrication of the principle of inflammability from the bodies into which it had entered as a constituent. This process is accomplished either by the introduction of fire, when the phlogiston is too firmly fixed and fettered, or by the intestine motion of the parts, as in fermentation and mixture, relying thus on internal and external causes. Animal heat, Leslie decided, is not like the process of fermentation that Franklin described. Fermentation "signifies both the smallest and the greatest degree of intestine motion of the particles of a fluid abounding with elastick and inflammable matter, from which new combinations are formed." In

[38] *Ibid.*, pp. 159–162, 173. Leslie cites *Experiments and Observations on Different Kinds of Air*, vol. III (London, 1777), p. 71. The paper, "Observations on Respiration, and the Use of the Blood," pp. 55–84, first appeared in *Phil. Trans. Roy. Soc. (London)*, 66 (1776), 226–248.

this sense, animal juices are in some degree of ferment, for they contain phlogiston and are always changing. And when reflecting on these changes and realizing that animal heat is not high, Leslie concluded "that the cause assigned is adequate to the effect, and sufficient to account for the whole phaenomena." Leslie was forced to admit, however, that it is only by assumption from the arguments presented that one can reach the conclusion that the phlogiston in animals is the sole cause of animal heat, for this cannot be directly demonstrated, but must be deduced.[39]

Leslie's theory emerges as something more than "merely an hypothesis"; it is a good argument. But just argument; he has not provided an experimental background of his own, but skips nimbly among the works of others. Leslie did not really develop any of the "radical principles" which he promised, but rather presented a hypothesis which blends old and new—fermentation and intestine motion emerging alongside of fixed air and phlogiston. He has really avoided the question of airs and gases, for he used the new discussion of them merely to prove that phlogiston was evolved. His proposal is qualitative and he makes no attempt to correlate the quantity of phlogiston with the amount of heat. In fact, he made no attempt to ascertain the quantity of either. He had already pointed out that according to his theory the degree of flammability was not even directly proportional to the amount of phlogiston, but was constantly altered depending upon the substance with which the phlogiston was in combination.[40] Although surely aware of the distinction,[41] he has seemingly made no attempt to demonstrate the relation between the quantity of heat pro-

[39] Leslie, *Philosophical Inquiry*, pp. 219, 239, 243, 258–260, 255–256.
[40] *Ibid.*, pp. 90, 197–198.
[41] He has a discussion at the close of the book relating the theory of latent heat to the cooling of bodies through evaporation and perspiration; *ibid.*, pp. 313f. 326f.

duced by the animal and its temperature. He has, however, avoided one problem that much confounded his contemporaries, the site of heat production. The consequence of placing the evolution of phlogiston in the blood as it passes through the extremities is that the heat is evolved throughout the animal system and not in the lungs alone, where it would necessitate some special mechanism to avoid overheating. In the same instance, however, Leslie found he had to rely on what were surely dubious measurements purporting to show that the venous blood is warmer than the arterial.

Leslie's treatise makes interesting reading, if for no other reason than that it was published at just the time that work was in progress on two quite radical proposals designed to account for animal heat, those of Adair Crawford and Antoine Lavoisier.

Crawford, like Leslie, had studied at the Scottish Universities.[42] In his first published work[43] he acknowledged his indebtedness to two of his teachers. He related that he attended the lectures of Dr. Irvine[44] at Glasgow and continued:

[42] See my bibliobiographical study of Adair Crawford.

[43] Adair Crawford, *Experiments and Observations on Animal Heat, and the Inflammation of Combustible Bodies* (London, 1779).

[44] William Irvine, 1743–1787, Lecturer on Materia Medica, 1766, Professor of Chemistry, 1770, Glasgow. Crawford's debt to Irvine may be greater than it appears on the surface but the debt was clearly acknowledged and it seems extreme to accuse Crawford of having "plagiarized" the work of Irvine, as Andrew Kent has done: *An Eighteenth Century Lectureship in Chemistry. Essays and Bicentenary Addresses Relating to the Chemistry Department (1747) of Glasgow University (1451),* ed. Andrew Kent (Glasgow: Jackson, 1950), p. 147. Irvine's theory of latent heat was not published by him during his lifetime, or for that matter during the formative period of heat theory. It appeared first in "A Letter from Mr. Irvine Concerning the Late Dr. Irvine, of Glasgow, His Doctrine, which Ascribes the Disappearance of Heat, without Increase of Temperature, to a Change of Capacity in Bodies, and that of Dr. Black, which Supposes Caloric to Become Latent by Chemical Combination with Bodies; with Particular Remarks

It is a tribute of justice which I owe to this philosopher, to acknowledge, that the solution which he has given of Dr. Black's celebrated discovery of latent heat, or of the heat which is produced by the congelation of water, spermaceti, bees wax, metals, &c. suggested the views which give rise to my experiments.[45]

Irvine had concluded that there was a difference between the absolute heat[46] of fixed and atmospherical airs. Crawford wished to determine experimentally in which state air had the greater quantity of absolute heat.[47] He had obviously made the connection between the alteration of atmospheric to fixed air in the lungs of animals and the possibility of this as a source of animal heat. For he immediately added his observation about the process in which warm animals are continually diffusing heat to the surrounding medium and his speculation that there must be a proportional supply to repair the waste. The animal, he concluded, must contain some power of exciting or collecting heat. He relates that "with a view to discover the nature of this power, I made

on the Mistakes of Dr. Thompson, in His Accounts of these Doctrines," *J. Nat. Phil., Chem., Arts* (Nicholson's), 6 (1803), 25–31; and later at greater length in William Irvine, *Essays, Chiefly on Chemical Subjects*, ed. and introd. by his son William Irvine (London, 1805).

Among the *Essays* of the posthumously published volume is one entitled "On the Effects of Heat and Cold on Animal Bodies," read in 1769 to the Literary Society of Glasgow. Although Irvine does not propose a theory of animal heat or really review the theories of others, he poses several questions which were surely calculated to develop further interest. He intimated that "nervous energy" might be the crucial factor in the preservation of animal life (pp. 196–197).

[45] Crawford, *Animal Heat* (1779), p. 17, n.

[46] Crawford notes that bodies differ in their capacities for absolute heat (or quantity of heat).

[47] Crawford, *Animal Heat*, p. 17, n. See William Cleghorn, *Disputatio Physica Inauguralis, Theoriam Ignis complectens* (Edinburgh, 1779), translated and edited by Douglas McKie and Niels H. deV. Heathcote as "William Cleghorn's 'De Igne' (1779), *Ann. Sci.*, 14 (1958), 1–82. Cleghorn, a student of Black's at Edinburgh, credits Crawford with making the quantitative determinations of the amount of heat or fire in the two kinds of air (pp. 31, 33, 35): "Crawford has established the facts."

a variety of experiments, in the summer of 1777."[48] These experiments, carried out in Glasgow, are the ones that Crawford communicated to Drs. Reid[49] and Irvine and Mr. Wilson.[50]

Crawford's study of animal heat is markedly different from those pursued by most of his predecessors. Although clearly based on a hypothesis, it relies to a much greater extent than had previous theories upon the experimental data brought forward. Making use of the method of mixtures, which Crawford must have heard much of during his years at Glasgow, he carried out a series of experiments in which he determined the "absolute heats" of several substances, including blood, by comparing them with water.[51] Noting that he found in blood a remarkable accumulation of heat, he claimed that he was led "to suspect, that it absorbs heat from the air, in the process of respiration." Crawford strengthened his supposition with two already common observations: one, that animals with lungs, and thus able to inspire great amounts of air, are warm, while animals without respiratory organs remain at about the same temperature as the environment; secondly, that the warmest animals

[48] Crawford, *Animal Heat* (1779), p. 18.

[49] Probably Thomas Reid, 1710–1796, Professor of Moral Philosophy, Glasgow, 1764–1796, under whom Crawford matriculated in 1764.

[50] Probably Patrick Wilson, 1743–1811, a fellow student at Glasgow, and the son of Prof. Alexander Wilson, whom he succeeded as Professor of Astronomy in 1784. Patrick Wilson's interest in problems of heat and cold is evidenced in some published papers; see "An Account of a Most Extraordinary Degree of Cold at Glasgow in January Last . . . ," *Phil. Trans. Roy. Soc. (London)*, 70 (1780), 451–473; 71 (1781), 386–394. Crawford, *Experiments and Observations on Animal Heat, and the Inflammation of Combustible Bodies* (London, 1788, 2nd edition with very large additions), advertisement.

[51] Crawford, *Animal Heat* (1779), pp. 20ff. See also McKie and Heathcote, *Specific and Latent Heats*, pp. 126–128, for a discussion of Crawford's use of mixtures, his assumption of water as a standard, and his corrections for heat lost to the vessel.

are those whose respiratory organs are largest in proportion to their over-all bulk.[52] To these he added the observation, probably derived from Joseph Black, that in any one animal the degree of heat is proportional to the quantity of air inspired in a given time and that consequently animal heat is increased by exercise or anything else that accelerates the respiration.[53] In this last comment Crawford has taken the common assertion that heat increases with exercise and related it to the quantity of air respired rather than to increased attrition of the blood in which case the inspired air had been seen as a cooling agent. The same data, it would seem, may serve many ends.

In Crawford's hands these observations were formulated into three propositions which taken together are the bases of his theory of animal heat. First, he proposed to demonstrate that atmospheric air contains more absolute heat than the same air after expiration from the lungs. Second, he would show that the blood in the pulmonary vein (arterial blood) after its passage through the lungs contains more absolute heat than the blood passing from the heart to the lungs through the pulmonary artery (venous blood). Lastly, he believed he could prove that the capacity of a body for heat is reduced by the addition of phlogiston and increased by the separation of phlogiston.[54] The first two propositions rely directly on Joseph Black's theory of the capacity for heat, but utilize the idea in a way that Black seemed not to have done. Number one seems to have been suggested directly by Irvine. Both proposals are clearly capable of direct experimental verification, using the method of mixtures that Black had developed. These two proposi-

[52] Crawford, *Animal Heat* (1779), pp. 30–31. Cleghorn repeats these observations almost in the words used by Crawford, "**Power of Producing Cold**," p. 33.

[53] Crawford, *Animal Heat* (1779), p. 31.

[54] *Ibid.*, pp. 31–32, 54, 58.

tions clearly indicate that Crawford had distinguished amount of heat from temperature. The third proposition was probably derived from Irvine, who had shown already that in chemical combinations the capacities for heat of bodies are altered.[55]

Crawford cited Priestley as his authority on the alteration of air during respiration. Atmospheric air was converted into fixed air (forms precipitate in lime water) and phlogisticated air (no precipitate, extinguishes flames, noxious to animals). With this in mind, Crawford turned to a determination of the accuracy of his first proposition by measuring the absolute heats of atmospheric, fixed, and phlogisticated air. These experiments, probably the earliest made to determine the specific heats of gases, utilized bladders, filled with the gas and then immersed in water. He was able to demonstrate, to his own satisfaction, that the "absolute heat" of atmospheric air is greater than that of either fixed or phlogisticated air. He conducted further experiments which showed that dephlogisticated air (obtained from red lead) has an "absolute heat" in the proportion of 4.6 : 1 to that of atmospheric air. This last is important because Crawford had claimed as a corollary of his proposition that the amount of absolute heat in a respirable air was in proportion to its purity or ability to support life, and Priestley had demonstrated that dephlogisticated air supported life five times as well as atmospheric air.[56]

The second proposition was proved in a similar manner

[55] See William Irvine, *Essays*, pp. 151, 180.

[56] Crawford, *Animal Heat* (1779), pp. 32, 34, 42, 53. Referring back to these experiments in the second edition (1788), Crawford comments: "As the trials recited in that publication had been made under many disadvantages, I soon afterwards found, upon a careful repetition of them, that I had fallen into considerable mistakes in my conclusions respecting the quantities of heat contained in the permanently elastic fluids." The differences in heat capacity were found to be great enough, however, not to effect his explanations (advertisement).

and yielded the results that the "absolute heat" of venous blood in proportion to arterial blood was as $10 : 11\frac{1}{2}$.[57] This distinction was of great importance and was much scrutinized in the years following. Crawford no longer had to claim that the temperature of arterial blood is greater, but could now assert instead that it holds more heat while remaining at essentially the same temperature.

The third proposition was proved using metals and their calces in mixture experiments with water. It was found that calces, from which phlogiston has been separated, have a higher heat capacity. "Heat, therefore, and phlogiston," said Crawford

appear to be two opposite principles in nature. By the action of heat upon bodies, the force of their attraction to phlogiston is diminished; and by the action of phlogiston, a part of the absolute heat, which exists in all bodies as an elementary principle, is expelled.[58]

Thus in proposition three a mode of heat transfer is developed which can work independently of temperature changes.

"Animal heat," Crawford commented, "seems to depend upon a process, similar to a chemical elective attraction." The atmospheric air that is drawn into the lungs has a great amount of "absolute heat," while the blood returning from the extremities is impregnated with phlogiston. The attraction of the air for the phlogiston, Crawford claimed, is greater than that of the blood for phlogiston; consequently, the phlogiston leaves the blood and combines with the air. In combining with the phlogiston, the air is obliged to give up some of its absolute heat while at the same time the blood on being separated from phlogiston has its capacity for heat

[57] Crawford, *Animal Heat* (1779), p. 58.
[58] *Ibid.*, pp. 59, 68.

increased and thus combines with the heat detached from the air.[59]

Priestley had shown, according to Crawford, that arterial blood has a strong attraction for phlogiston; thus during its circulation through the extremities the arterial blood will imbibe the phlogistic principle from those parts which retain it with least force and consequently the venous blood on returning to the lungs will be saturated with phlogiston. Therein lies the source of animal heat. "In proportion, therefore," Crawford said,

as the blood which had been dephlogisticated by the process of respiration, becomes again combined with phlogiston, in the course of the circulation, it will gradually give out that heat which it had received in the lungs, and diffuse it over the whole system.

But the extremities in discharging phlogiston have their capacity for heat increased and thus "absorb" the heat coming from the blood. If all the heat were absorbed then no part of the heat received in the lungs would become sensible. And here another weak link is exposed, for Crawford is forced to conclude that since "sensible heat" is produced by the circulation it either arises from an entirely independent source or in some way comes from the heat taken in at the lungs and absorbed by the blood. At this point, acting in the "true" scientific spirit, Crawford introduced a rule of simplicity, by which he avoided adding multiple causes. He was able to conclude that, although some heat from the

---

[59] *Ibid.*, p. 73. This system of double elective attraction has been schematized by Partington and McKie, "Phlogiston Theory.—III," p. 349:

atmospheric air + venous blood → phlogisticated air + arterial blood
or,

$$(air + heat) + (blood + phlogiston) \rightarrow (air + phlogiston) + (blood + heat)$$

i.e.,

$$AB + CD \rightarrow AD + CB.$$

blood is absorbed by the parts, its proportion is inconsiderable by comparison with that becoming sensible during the circulation.[60]

Another weak point in Crawford's system is the theory of attraction itself. He seemed able to assume that it works, in part by relying on the testimony of Priestley regarding the attraction in arterial blood for phlogiston, and in part by its consistency in providing for heat transfer within the system he has outlined.[61] Within this framework he was forced to treat heat as a substance and, indeed, argued mildly for this view in the later pages of the treatise.[62]

Crawford has avoided the problem of the overheated lungs, which was a difficulty for most early respiration theories of animal heat, by having the evolution of the "sensible" heat take place in the extremities at the time of the blood's imbibing of phlogiston.

Having developed his theory of animal heat, Crawford turned to combustion, saying,

I shall hereafter show, that the heat which is produced by this process, is similar to that which is produced by the inflammation of combustible bodies, with this difference, that, in the latter instance, the fire is separated from the air, in the former, from the blood.[63]

[60] Crawford, *Animal Heat* (1779), pp. 73–75.

[61] Crawford here seems to lean toward Black's theory, in which heat, itself a substance, is essentially chemically combined with other substances. Irvine had been opposed to the view that heat was material. See Kent, *Eighteenth Century Lectureship*, p. 144, and Irvine, "A Letter," *passim.*

[62] Crawford, *Animal Heat* (1779), p. 116. He is not as happy with this view in the second edition (1788), p. 363. Cleghorn, whom McKie and Heathcote credit with providing the first detailed exposition of the material theory of heat, adopts Crawford's explanation of animal heat and in his report of it treats heat as material in nature; Cleghorn, *De Igne,* pp. 1, 33, 35.

[63] Crawford, *Animal Heat* (1779), p. 76.

The initial demonstration of the similarity of respiration and combustion had been provided by Black when he showed that both processes cause a similar alteration of the air. Black had further speculated that animal heat was derived from the same source as the heat of combustion. Crawford provided experimental links between respiration and animal heat and with them a consistent theory. It is probable that for Black the idea of animal heat came as an addendum to his general views on combustion; the reverse seems to be the case for Crawford.

The heat of inflammation, Crawford believed, comes from the air (as in respiration), while phlogiston is separated from the inflammable body. Again an attraction takes place, the phlogiston combines with the air, and the latter gives off a great proportion of absolute heat, which, suddenly extracted, bursts into flame, producing sensible heat.[64]

After pointing to a number of phenomena in which heat is separated from the air by phlogiston (as when common air is mixed with nitrous air, exploded with inflammable air, or diminished and rendered noxious by putrefaction, combustion, or an electric spark), Crawford concluded:

It appears, upon the whole, that atmospherical air contains, in its composition, a great quantity of fire or of absolute heat. By the separation of a portion of this fire in the lungs, it supports the temperature of the arterial blood, and thus communicates that *pabulum vitae*, which is so essential to the preservation of the animal kingdom. And, finally, by a similar process, it maintains those natural and artificial fires which are excited by the inflammation of combustible bodies.[65]

This was the extent of Adair Crawford's work on animal

[64] *Ibid.*, p. 77.
[65] *Ibid.*, pp. 78–79.

heat in 1779,[66] although he did include two further observations which were to become important in the much-revised edition of 1788. First, he reported an experiment which shows that the quantity of air phlogisticated by a man in a minute was the same as that altered by a candle. He concluded that the man, therefore, must have derived as much heat from the air as was produced by the burning candle. He also reported having made a number of experiments in which he measured the quantity of air phlogisticated by the calcination of iron and found it to be approximately equal to the quantity of metal calcined, "from which," he said, "we may calculate the heat produced by the process." This statement, he believed, could be generalized to say that the quantity of fire separated from the air will be in proportion to the quantity of phlogiston combined with it in a given period of time.[67] Thus Crawford has outlined a type of indirect calorimetry. What he has not yet attained is the method of direct calorimetry he utilized for the 1788 edition and which was crucial for the study of animal heat prior to the determination of heats of combustion of the various gases.[68] As yet no one had directly measured the amount of heat produced by an animal.

It was this early, imperfect work of Crawford's which brought him immediate notice. The inaccuracies of reporting of Black's discoveries[69] notwithstanding, Crawford was

[66] His explanations of several important phenomena of animal heat have not been discussed here, as they do not alter the understanding of his theory, although they lend support to it.

[67] Crawford, *Animal Heat* (1779), pp. 80, 107, 108–109. The calculations are made at length.

[68] The heats of combustion themselves were first measured by Crawford using a form of direct calorimetry.

[69] See McKie and Heathcote, *Specific and Latent Heats*, pp. 38ff, for a full indictment. It is interesting that Black in the posthumously published lectures gives no indication that he was mistreated at the hands of Crawford, but instead cites Crawford's treatise rather favorably. The

widely credited with developing a new theory of animal heat and of combustion.[70] His work was extensively reported in France,[71] and appeared in German and Italian translation.[72]

McKie and Heathcote charges are somewhat overstated; speaking of Crawford's book, for example, they say: "This work, highly personal in tone, gave detailed accounts of the author's own researches, and the unwary reader, remote from the scene of events, might well imagine on perusing Crawford's pages that he was being introduced, if not to one of the founders of the quantitative science of heat, at least to one of its extensive improvers" (p. 38). Later on they refer to him as "a muddled and careless writer on matters of history" (p. 40). All this, despite the fact that Crawford had acknowledged his debt to both Black and Irvine, had indeed provided one of the first extensive accounts of the new heat theory, and had added to his account experimentation of his own, including many firsts, such as an attempt to determine the specific heat of gases.

I am not quite sure where the denigration of Crawford's work began, but it may well have been in the circle of friends of Joseph Black, although excluding Black himself. James Watt writing to Adam Ferguson, Feb. 8, 1801, said: "Dr. Crawford seems not to have been much disposed to do either Dr. Black or Dr. Irvine justice, and consequently was willing to deprive either of them of the honour of discoveries which were the foundation, and indeed the only valuable parts, of his own voluminous writings on the subject. His own theories and experiments are, in my opinion, of little value, although they serve as one of the props to the French theories" (Muirhead, *James Watt*, p. 275). Perhaps the last comment is the key to some of the early anti-Crawford attitudes. See my bibliobiographical study of Crawford for an analysis of Crawford's role in transmitting the new theory of heat.

[70] Alfred Novak, "Ideas on Respiration—and Adair Crawford," *Janus*, 67 (1958), 180–197, is quite wrong when he suggests (p. 196) that Crawford's work went unnoticed because the conceptual scheme was unattractive.

[71] See Jean Hyacinth Magellan, "Extrait d'une Lettre de M. Magellan, de la Société Royale de Londres, sur les Montres nouvelles qui n'ont pas besoin d'être montées, sur celle de M. Mudge & sur l'Ouvrage de M. Crawford," *Observations sur la Physique*, 16 (1780), 60–63; "Lettre de M. Magellan, à l'Auteur de ce Journal, sur le Mémoire suivant," *Obs. Phys.*, 17 (1781), 369–375; "Essai sur la nouvelle Théorie du Feu élémentaire, & de la Chaleur des Corps," *Obs. Phys.*, 17 (1781), 375–386, 411–422.

[72] *Versuche und Beobachtungen über die thierische Wärme und die Entzundung brennender körper* (Leipzig, 1785, 1789), translated by Crell; and *Sperienze ed osservazione sul calore animale e sull' inflam-*

In 1777 Antoine Lavoisier read two papers in which he put forward his first hypotheses on the source of animal heat.[73] The papers deal with the phenomena which occur in respiration and combustion; they recognize the similarity of the processes involved and speculatively conclude that respiration as a type of combustion should be capable of generating the heat of animals.

Lavoisier noted, from experiment, that five-sixths of the air breathed by an animal is incapable of supporting either respiration or combustion and that only one-sixth[74] is respirable. The same portion of the air, called "dephlogisticated" by Priestley, and "pure" by Lavoisier, not only supports the vital function of respiration, but also is necessary for the calcining of metals and for combustion. During respiration and combustion this "pure" or "respirable" portion of the air is converted into "acide crayeux aériform" (fixed air).[75] Lavoisier had been carrying out experiments on birds and guinea pigs which he permitted to respire a limited amount of air under a bell jar.[76] He proposed two

mazione de' corpi combustibili, in _Opus coli scelti sulle scienze e sulle arti,_ ed. Carlo Amoretti and Francesco Soave, 22 vols. (Milan, 1778–1803), vol. 3. Reviews of Crawford's work were numerous both in England and on the continent.

[73] A. L. Lavoisier, "Expériences sur la Respiration des Animaux, et sur les Changements qui Arrivent à l'Air en Passant par leur Poumon," _Mém. Acad. Roy. Sci.,_ 1777 (1780), 185–194; "Mémoire sur la Combustion en Général," _ibid.,_ 592–600. Although the papers were read in 1777 and included in the _Mémoires_ for that year they were actually not published until 1780. As late as December 1779 Lavoisier was still reading a part of his "Mémoire sur la Combustion" of 1777. See the report of the _Procès-Verbaux_ of the Academy of Science in Maurice Daumas, _Lavoisier, Théoricien et Expérimentateur_ (Paris: Presses Universitaires de France, 1955), p. 44.

[74] In the paper it reads one-fifth, obviously a typographical error, Lavoisier, "Expériences sur la Respiration," p. 188.

[75] _Ibid.,_ pp. 188–190. "Mémoire sur la Combustion," pp. 593, 598, 599.

[76] See Daumas, _Lavoisier,_ pp. 37–38, for entries from laboratory note books of Lavoisier. Daumas also records from the _Procès-Verbaux_ of

alternative explanations of the procedure by which the air was altered during the respiration of these animals. The first suggested that the respirable part of the air is converted into fixed air (*acide crayeux aériform*) in the lungs themselves as it passes through them during respiration. The second proposed that in the lungs an exchange occurs, the respirable air being absorbed while the lungs release a nearly similar volume of fixed air. For support of the first hypothesis Lavoisier turned to work he had done in 1775 in which he had shown that highly respirable air had been converted to fixed air by the burning of powdered charcoal. It is possible, he suggested, that respiration is capable of producing the same process in the lungs, and thus altering respirable air into fixed air.[77]

Lavoisier sought support for his second hypothesis in an interesting analogy. The air, he pointed out, causes a redness in certain bodies, for example, mercury, lead, and iron, when it combines with them. Since calcination of metals and respiration are similar in requiring the same portion of the air, perhaps they are also similar in the color of their residues. May we not suppose, Lavoisier suggested, that the red color of blood depends upon its combination with respirable air?[78] Lavoisier had given more consideration to this problem than is evidenced by his simple analogy. Just

---

the Academy an entry indicating that Lavoisier was prepared to read a paper "Sur la décomposition de l'air dans le poumon," as early as November 13, 1776, *ibid.*, p. 37. M. Berthelot, *La Révolution Chimique, Lavoisier* (Paris, 1890), pp. 290–291, records for 13 October 1776 "expériences faites à Montigny par MM. Trudaine et Lavoisier" involving the respiration of birds.

[77] Lavoisier, "Expériences sur la Respiration," p. 191.

[78] *Ibid.*, p. 192. Lavoisier notes that Priestley and Cigna had come to a different conclusion concerning the source of the redness of blood. See Priestley, "Observations on Respiration," pp. 230, 234, and Cigna, "De colore Sanguinis," p. 73.

prior to reading his paper on respiration, he had submitted a "Mémoire sur les changements que le sang éprouve dans les poumons et sur le mécanisme de la respiration."[79] That Lavoisier was not wholly satisfied with his results might be inferred from the fact that the paper remained unpublished.[80]

At this point, and for some time after, Lavoisier wished to believe the truth of both hypotheses: respirable air was altered in the lungs and also absorbed in them. He closed the paper on respiration with mention of further experiments he had carried out but had no room to present in the current memoir, and suggested that they would throw more light on respiration and combustion, processes which, he felt, are more closely related than is immediately apparent.[81]

Although claiming that not until a later memoir would he show the analogy that exists between the respiration of animals, combustion, and calcination, Lavoisier provided a clear indication of his line of thought in his 1777 "Mémoire sur la combustion." Having already indicated that in both respiration and combustion there is an alteration of "pure air," he now demonstrated how this alteration might prove to be the source of the heat evolved. All types of air, as he had shown in another paper, were a combination of some substance (or base) with the matter of fire or light.[82] In a system very much like that described by Crawford, La-

[79] This was read April 9, 1777. See the report of the *Procès-Verbaux* of the Academy in Daumas, *Lavoisier*, p. 38.

[80] It was included as part of the paper that was read to the Société Royal de Médecine in 1785, "Mémoire sur les altérations qui arrivent à l'air dans plusieurs circonstances où se trouvent les hommes réunis en société," *Hist. Soc. R. Médec.*, 1782–83 [1787], pp. 569–582.

[81] Lavoisier, "Expériences sur la Respiration," pp. 193, 194.

[82] "De la combinaison de la Matière du feu avec les Fluides Evaporables, et de la formation des Fluides élastiques aëriformes," *Mém. Acad. Sci.*, 1777 [1780], pp. 420–432.

voisier saw pure air (Crawford's dephlogisticated air) as an igneous combination of the matter of fire or light (acting as a dissolvent) and another substance (as a base). If the base is brought into contact with a substance for which it has more affinity, it unites instantly with it, and the dissolvent is thus freed and escapes as flame, heat, or light.[83]

Carbon and carbonaceous materials are seen to act in this manner, combining with the base of pure air and forming an acid, "fixed air" or "*acide crayeux,*" thus freeing the matter of fire. It is clear from this theory of combustion why there is no inflammation either in a vacuum or in any aeriform combination in which the matter of fire has a very great affinity with the base with which it is combined.[84]

Since he had shown in the memoir on respiration that pure air in passing through the lungs is converted in part to fixed air and that a similar decomposition occurs in the combustion of charcoal, if fire is evolved in the latter process it should likewise be expected in respiration. This,

[83] Lavoisier, "Mémoire sur la Combustion," p. 596. Henry Guerlac, "A Lost Memoir of Lavoisier," *Isis,* 50 (1959), 128–129, suggests that Lavoisier had developed a major part of his theory of caloric as early as 1772. The paper in which these views might have been spelled out, "Mémoire sur le feu élémentaire," seems not to have been published during Lavoisier's lifetime. M. René Fric, the editor of the Lavoisier *Correspondance,* has recently discovered this paper, bearing the title "Réflexions sur la Combinaison de la Matière du Feu dans les Corps" and published it in his "Contribution à l'étude de l'évolution des idées de Lavoisier sur la nature de l'air et sur la calcination des métaux," *Arch. Internat. Hist. Sci.,* No. 47 (1959), pp. 137–168. In his new study of Lavoisier's early work on combustion Henry Guerlac draws the identity between these two Lavoisier papers and notes a striking resemblance between the theory of heat enunciated by Lavoisier and that found in the work of Joseph Black. Important for our study is the implication already visible in Lavoisier's work of 1772 that heat can be combined with another substance without affecting the thermometer. Guerlac, *Lavoisier—The Crucial Year. The Background and Origin of His First Experiments on Combustion in 1772* (Ithaca, N.Y.: Cornell University Press, 1961), pp. 96–97.

[84] *Ibid.,* pp. 597, 599.

Lavoisier believed, occurred in the lungs during the interval between inhalation and exhalation. And it is this matter of fire which is distributed with the blood throughout the animal economy and where it serves to maintain the constant animal heat of about 32½ degrees Réaumur. Although quick to admit that the theory is speculative, Lavoisier believed it was founded on two certain and incontestable facts: (1) the decomposition of air in the lungs, and (2) the evolution of the matter of fire which accompanies all decomposition of pure air, resulting in fixed air. Lavoisier found further support, as did Crawford, in the common observation that in nature it is traditionally the animals which respire that are warm-blooded and that this warmth increases as respiration becomes more frequent. He thus pointed to a constant relation between the warmth of an animal and the quantity of air entering and being decomposed in the lungs.[85]

Conceptually, the schemes proposed by Lavoisier and Crawford are quite similar to each other and different from that of Leslie. The work of all three men falls within the same time span, 1775–1780, but only Lavoisier and Crawford seem to have proposed a radically new source for animal heat and combustion. Crawford, although presenting his discussion in the language of the phlogiston theory (he was an Englishman and a friend of Priestley) does not seem hampered by it. Like Lavoisier, he talked in terms of a material heat or fire which can enter into combination with other substances through the power of elective attraction. Although his measurements are of dubious accuracy, Crawford, by 1779, had provided an "experimental" proof for the assertion that dephlogisticated (or pure) air contained more heat than fixed air. Lavoisier asserted that the heat of

[85] *Ibid.*, pp. 599–600.

animals was generated in the lungs[86] and at the same time provided the source of doubt by having the "pure air" absorbed by the blood.[87] Crawford successfully avoided the problem by his finding that the arterial blood had a greater capacity for heat than the venous blood and therefore could take up or absorb heat in its passage through the lungs. For both men animal heat had its source in the respiratory process which was at one with all other forms of combustion.

[86] *Ibid.*, p. 599.
[87] Lavoisier, "Expériences sur la Respiration," p. 193.

· VI ·

From Analogy to Chemistry

Whether or not we are ready to join one contemporary commentator and claim that Adair Crawford was "the first who attempted to ascertain, by direct experiment, the cause of animal heat,"[1] it is apparent that his initial work and the stimulus behind it encouraged many others to attempt experimental disproof, or verification of his theory. A glance at the proliferation of animal-heat literature during the last two decades of the eighteenth century makes this point clear.

John Elliott,[2] who published an essay on animal heat in 1780, just one year after Crawford's first volume, provides an interesting case in point. In an appendix to his own paper, Elliott relates that he came upon the books of Leslie and Crawford after completing his own work.[3] He, too, disagrees with Leslie, but is unbounded in his praise for Crawford, a praise which he considered all the more meaningful since he himself had been writing on the same subject and, like Crawford, had advanced several steps further than others. But this work of Elliott's stands in sharp contrast to

[1] W. B. Johnson, *History of the Progress and Present State of Animal Chemistry* (London, 1803), III, 85.

[2] See J. R. Partington and Douglas McKie, for the corrected biography of John Elliott, "Sir John Eliot, Bart. (1736–86), and John Elliot (1747–87)," *Ann. Sci.*, 6 (1948–50), 262–267.

[3] John Elliott, *Philosophical Observations on the Senses of Vision and Hearing; To which are Added, a Treatise on Harmonic Sounds, and an Essay on Combustion and Animal Heat* (London, 1780), pp. 205–206.

Crawford's, even though they are alike in point of view. The contrast is one that Elliott himself recognized; he related that he had not been in a position to carry out experiments and instead argued from facts that others had published and from analogy. None the less he claimed to have discovered, as had Crawford, that "combustion is a truly chymical process, and that it depends on the superior affinity or attraction between phlogiston and air."[4] Animal heat was treated as arising from a process analogous to combustion.

A brief scrutiny of Elliott's theory is instructive, for it illustrates by contrast the originality of Crawford and Lavoisier, by showing which parts of the extant tradition when utilized led to the development of successful theories. Priestley's observations on the alteration of "airs" that occurred during respiration and combustion formed the basis of Elliott's speculations. The suggestion made by Priestley that respiration served to carry off the phlogiston which the blood acquired during circulation confirmed in Elliott's view the close analogy between respiration and combustion.[5] Priestley's work, as was pointed out above, had a seminal role in the efforts of both Crawford and Lavoisier, and also had been appealed to by Leslie in his attempt to develop a theory of animal heat.

Elliott's ideas about what happens during respiration are not unlike Crawford's. He claimed to have demonstrated that heat was evolved whenever phlogiston combined with air,[6] and conjectured on this basis that phlogiston by combining with air weakened the attraction that air had for

---

[4] *Ibid.*, pp. 207–208.

[5] *Ibid.*, p. 184.

[6] This, it will be noted, is quite different from the view of Leslie, who believed that heat was generated whenever phlogiston was evolved from a substance with which it was combined. It was the air which was combined with fire or heat in the systems of Crawford and Lavoisier.

fire and consequently the fire gravitated toward the surrounding bodies. Thus, Elliott believed that the blood, on returning to the left ventricle of the heart from the lungs, should be hotter than the blood which entered the lungs. It is this excess of heat which he believed accounted for the warmth of the animal body. The reason that the blood was able to take up the extra heat, Elliott conjectured, was that its power of attraction for fire was increased when the particles of blood gave up phlogiston. Elliott has proposed a scheme of double attraction almost identical with the one outlined by Crawford. During the systemic circulation more heat is evolved because the attraction of the blood for fire is lessened when phlogiston is imparted to it by the nerves. "Heat therefore," Elliott pointed out, "will follow for the same reason that it follows on the combination of phlogiston with air . . . only in a less degree."[7]

The element missing from Elliott's proposal, and so effectively utilized by Crawford, was the new theory of specific heat (or capacity for heat as used by Crawford). Upon reading Crawford, Elliott gave partial recognition of this important addition to the theory of animal heat, for he said, "I had fallen into an error in imagining that the blood is partly heated in the lungs." However, Elliott refused to give up his belief that heat is generated in the lungs upon the decomposition of air, and in support of this idea he turned to the old "common-sense" observation that air upon expiration from the lungs is hotter than when inspired. What really happens in the lungs, Elliott insisted, is that the evaporation which occurs in this organ is so proportioned that the heat of the blood is not increased in its

[7] *Ibid.*, pp. 186–188. The role of the nerves as the secretors of phlogiston is different in detail, but conceptually similar to Crawford's view of phlogiston evolution in the system.

pulmonary passage.[8] Elliott, by his inability to deal successfully with the heat he believed to have been separated from the air, has clearly demonstrated the importance of the new heat theory as used by Crawford. The change in capacity for heat which Crawford believed to take place in the blood during its passage through the lungs obviated the difficulties encountered by the generation of heat in the lungs and the consequent problems involved in its distribution through the body.

That the early proposals of Crawford and Lavoisier were not sufficiently compelling to be considered as the answer to the question of the origin of animal heat is attested to by several treatises which ignored or contradicted the attempts of these two authors. Edward Rigby, an English physician, came to the problem with about the same background as John Elliott and yet proposed an answer which was distinctly different. Rigby had read Priestley's memoir on respiration and also had knowledge, if not understanding, of the theory of latent heat.[9]

The standard sources of heat are inadequate, Rigby claimed, to account for animal heat and he argued that there must be some *latent* and internal source available to provide the observed warmth. Rigby interpreted latent heat as meaning that over and above the sensible heat of bodies there exists a good deal "of the matter of heat in a concealed and fixed state, and which is not separated from them but by their decomposition."[10]

Turning to the theory of Franklin, which envisaged a "fluid fire" as being attracted by and becoming a part of

---

[8] *Ibid.*, pp. 213–215.
[9] Edward Rigby, *An Essay on the Theory of the Production of Animal Heat, and on its Application in the Treatment of Cutaneous Eruptions, Inflammations, and Some Other Diseases* (London, 1785), pp. 2–5, 13.
[10] *Ibid.*, pp. 3–4.

the substance of plants, Rigby proposed that heat is produced in the stomach during the decomposition of the foods, with consequent freeing of the "fluid fire." Priestley, he noted, recovered inflammable air from meat and, if one accepts Rigby's suggestion that inflammable air is a combination of phlogiston and the matter of heat, here is further proof of his hypothesis.[11]

No experimental evidence was offered with this new proposal to account for animal heat, and Rigby, unlike Elliott, made no apology for the lack of empirical data or in any way indicated that he thought any such support to be necessary. He presented his argument from tradition and analogy. For example, Rigby cited the supposed closeness of the matter of heat to oily and fatty substances, and suggested that the fatty parts of the body are probably the repository of excess amounts of the material heat.[12] A theory of the kind proposed by Rigby, while perhaps helpful to its author's understanding, passes into and out of the literature without affecting the work of others, even though it makes use of a body of knowledge common to many.

Crawford's hypothesis was of a different nature, however, and was immediately recognized for its experimental basis and the aid it might give to stimulating other researches. One contemporary reviewer, comparing the work of Leslie and Crawford, commented that "Dr. Leslie pursued his researches with argument and inductive reasoning, while Mr. Crawford has endeavoured to erect his hypothesis upon the basis of experiment."[13] Another commentator,

[11] *Ibid.*, pp. 11–14.
[12] *Ibid.*, pp. 20, 58.
[13] Review of Adair Crawford's *Experiments and Observations on Animal Heat, Critical Review, or, Annals of Literature*, 48 (1779), 182. See also *Medical Commentaries*, 6 (1779), 399–414, for a laudatory review of Crawford's books, which suggested that Crawford's work might open up new fields of endeavor.

William Morgan, an actuary, was not quite as kind to Crawford and pointed out that many of Crawford's conclusions were drawn on the basis of one or at most two experiments, a charge that Crawford freely admitted.[14] A reviewer of Morgan's book took up his attack against Crawford and claimed that "in a variety of foreign publications Dr. Crawford receives such honourable titles as are rarely prefixed to the names even of a Boyle or a Newton." Although expecting much auxiliary support to be brought forth backing Crawford, the anonymous reviewer found "that one of the first combatants who have taken the field, has entered the lists against him; and that, with an aspect truly formidable and menacing."[15]

Morgan's criticisms were mounted on several levels; on the one hand he was unhappy with Crawford's mathematical propositions and felt that subjects such as those with which Crawford dealt stood upon too precarious grounds to admit the use of mathematics; plain reasoning should have been sufficient.[16] On the other hand, Morgan claimed that he had constructed apparatus for the purpose of finding a difference in the absolute heats of air before and after respiration, but was unable to find any difference. Also, he did not believe that atmospheric air was ever converted to fixed air.[17] All this provided interesting polemic, but hardly touched the reputation of Crawford;[18] neither, for that

[14] William Morgan, *An Examination of Dr. Crawford's Theory of Heat and Combustion* (London, 1781).

[15] Review of William Morgan's *Examination of Dr. Crawford's Theory of Heat and Combustion, Critical Review, or, Annals of Literature,* 51 (1781), 212–213.

[16] Morgan, *An Examination,* p. 61. This is an interesting view for a man who was an actuarial mathematician.

[17] *Critical Review,* 51 (1781), 215.

[18] See, for example, William Nicholson, *A Dictionary of Chemistry* (London, 1795), I, 375ff, which refers to Crawford's work with high praise.

matter, did it substantially aid the study of the origin of animal heat.

The most serious challenge to Crawford's growing reputation as the founder of a new theory of animal heat came from Antoine Lavoisier. The early writings (prior to 1780) of Lavoisier dealing with animal heat do not propose a theory as acceptable, or provide experimental data as useful, as the contemporaneous work of Crawford. Beginning about 1780, however, Lavoisier and several associates carried out what is probably the most important group of experiments in the history of metabolic-heat studies. In the hands of Lavoisier animal heat was transposed from being a phenomenon *analogous* to some chemical process to the position of being treated itself as a chemical process. The germ of this change was clearly present in Lavoisier's early papers on respiration and combustion; the transition took place in a series of studies first by Lavoisier and Laplace and later by Lavoisier with the assistance of the physiologist Armand Séguin.

Lavoisier, when he began his studies of animal heat, had at his disposal almost all the same background information as was utilized by Crawford; and only because others with the same material to work with had failed to reach similar conclusions does it seem surprising that the theories as expounded by Lavoisier and Crawford were conceptually almost identical. In the September 1772 issue of Rozier's *Journal* there appeared an anonymous article, "Expériences du Docteur Black, sur la marche de la chaleur dans certain circonstances," which reported Black's discovery of latent heat.[19] The same periodical, in the number for June 1773,

[19] See Douglas McKie, "The 'Observations' of the Abbé François Rozier (1734–93)," *Ann. Sci.*, 13 (1957), 86. The text is reprinted. Henry Guerlac, "A Lost Memoir of Lavoisier," *Isis*, 50 (1959), 128–129, has demonstrated that Lavoisier was familiar with the report of Black's work on latent heat.

carried a French version of Daniel Rutherford's "Dissertation," the work referred to above as one of the earliest sources of Joseph Black's speculation on animal heat.[20] These articles were most probably read by Lavoisier. We also know that he closely followed the work of Joseph Priestley and Adair Crawford (of whose work extensive reports in French were available by 1781, the time at which Lavoisier was at work on his "Mémoire sur la Chaleur").[21]

In this important paper, written in collaboration with Pierre Simon de Laplace,[22] Lavoisier proposed a new theory of animal heat and provided the first direct calorimetric measurements of the heat generated by a living animal. The first part of their memoir deals at length with the general theory of heat and combustion and outlines their system

[20] See Douglas McKie, "Daniel Rutherford and the Discovery of Nitrogen," *Science Progress*, 29 (1935), 655.

[21] Antoine Lavoisier and Pierre Simon de Laplace, "Mémoire sur la Chaleur," *Mém. Acad. R. Sci.*, 1780 [1784], 355–408. This paper was first printed separately as *Mémoire sur la Chaleur. Lû à l'Académie Royale, le 28 Juin 1783* (Paris, 1783). Early in 1780 Jean Hyacinthe Magellan reported on Crawford's experiments and theory. See "Extrait d'une Lettre de M. Magellan, de la Société Royale de Londres, sur les Montres nouvelles qui n'ont pas besoin d'être montées, sur celle de M. Mudge et sur l'Ouvrage de M. Crawford," *Observations sur la Physique*, 16 (1780), 60–63.

[22] Just why the mathematician Laplace was involved in these studies is hard to tell. The first indication of collaboration between the two men came in Lavoisier's notebooks in 1778. See Maurice Daumas, *Lavoisier, Théoricien et Expérimentateur* (Paris: Presses Universitaires de France, 1955), pp. 41–42. Laplace was the actual reader of the memoir at the June 18, 1783 session of the Académie (*ibid.*, p. 48). Roger Hahn (personal communication, Feb. 8, 1959) informs me that to his knowledge this work on animal heat is Laplace's only mention of physiological problems. Hahn also informs me that Laplace was in correspondence with Sir Charles Blagden, an Englishman concerned with animal heat. See Charles Blagden, "Experiments and Observations in an heated Room," *Phil. Trans. Roy. Soc. (London)*, 65 (1775), 111–123, 484–494. Roger Hahn and Denis Duveen, "Deux lettres de Laplace à Lavoisier," *Rev. Hist. Sci.*, 11 (1958), 340, report a letter dealing with the mechanics of animal calorimetry, in which Laplace refers to "nostre théorie de la respiration," indicating that he considered the theory at least partly of his own invention.

of calorimetry. There is significance in the fact that they attached their discussion of animal heat to a broader discussion of the phenomena of heat in general. Firstly, this is the same course that Crawford took;[23] and secondly, it clearly implies that the authors saw no separation existing between the biological and physical-chemical worlds.

The hypothesis behind these experiments, claimed Lavoisier and Laplace, is essentially the one that Lavoisier had presented in the *Mémoires* for 1777 (and which is alluded to above).[24] Lavoisier had proposed that heat was evolved in any decomposition of "pure air," and that this process occurs in both combustion and respiration, producing "fixed air" as a residue. The heat evolved in respiration was suggested as the source of animal heat. The authors note the similarity of Crawford's explanation as published by him in 1779.[25] Even though the two publications were in agreement in regarding pure air as the source of heat in combustion and respiration, Lavoisier and Laplace pointed to what they considered to be an essential difference, the manner in which the heat and the pure air combined. Lavoisier and Laplace supposed a form of chemical combination between the air and the heat, while Crawford saw heat as existing more freely in the pure air, only disengaging itself when the pure air through combination with some other substance loses a part of its specific heat (or capacity for heat).[26] This difference, although affecting the mechanism of heat transfer, does not alter the conceptual simi-

---

[23] The very title of Crawford's book is suggestive of his outlook: *Experiments and Observations on Animal Heat, and the Inflammation of Combustible Bodies. Being an Attempt to Resolve these Phaenomena into a General Law of Nature.*

[24] See Chapter V. The paper in question was actually read in 1777 and published in 1780. Lavoisier and Laplace, "Sur la Chaleur," p. 394.

[25] *Ibid.*; Crawford's name is misspelled Craford.

[26] See Partington and McKie, "Historical Studies on the Phlogiston Theory.—III. Light and Heat in Combustion," *Ann. Sci.*, 3 (1938), 349.

larity between the proposals of Lavoisier and of Crawford.

In the present experiments, Lavoisier and Laplace wished to limit themselves to a comparison of the amounts of heat liberated in combustion and in respiration and a determination of the corresponding alterations of pure air. The apparatus used for the heat comparison was the ice calorimeter, a device of their own invention, which was based on the theoretical and practical advances made by Joseph Black in his work on the latent heat of fusion of ice. Burning coals or charcoal, or an animal, were placed in the receiver of the three-chambered calorimeter, the heat produced being calculated from the amount of ice melted.[27] The amount of fixed air produced, or of pure air altered by combustion and respiration, was determined by burning the coal, or allowing the animal to respire, under a bell jar or some other vessel which permitted collection of the respired air.[28] Taking possible errors into account, Lavoisier and Laplace claimed that their results showed that in the combustion of carbon and in the respiration of an animal the production of equal amounts of fixed air was accompanied by the generation of equal amounts of heat.[29]

Respiration, Lavoisier and Laplace concluded, is a slow

[27] The apparatus is described in detail in Lavoisier and Laplace, "Sur la Chaleur," pp. 369f and plates.

[28] *Ibid.*, pp. 395f. I will not scrutinize the computations involved here, but will mention two criticisms that were later leveled against the Lavoisier-Laplace procedure. First, the amount of water recovered from the melting ice was not a real measure of the ice melted, since much water remained fixed to the crushed ice. Second, an animal put in a chamber surrounded by ice was subjected to extreme conditions and further to conditions quite different from those in which the respiration alterations were measured. The authors indicated (p. 404) an awareness of the latter criticism.

[29] *Ibid.*, pp. 404–405. The authors presented their results with some hesitation, since their figures for the heat liberated in the combustion of carbon were based on only one experiment (p. 398). The guinea pig melted 13 ounces of ice in 10 hours, and only by allowing correction factors to account for 2½ ounces do they come within range of the theoretical 10.38 ounces which they expected (p. 405).

combustion which takes place in the interior portions of the lungs. No light or flame is observed, they claimed, because as soon as the matter of fire is released from the pure air it is absorbed by the humidity of the lungs. The heat is thence transferred to the blood which is in pulmonary transit and carried to the rest of the animal system. Air that the animal breathes was seen to serve two purposes: one, the removal of the base of fixed air from the blood, since an excess of it in the blood is injurious; two, the production, in the combination of the base of fixed air with the pure air, of a heat which serves to replace the heat that is constantly being lost to the environment.[30]

Thus far in the development of their theory Lavoisier and Laplace have not had recourse to one of the two ideas that played so important a role for Crawford and helped him avoid the difficulty of having an excess of heat freed in the lungs. Only when they turned to a discussion of the seeming equality in temperature in different parts of the body did Lavoisier and Laplace mention that the specific heat of blood is raised when it rids itself of the base of fixed air. They assumed that this occurred, without making reference to Crawford or to any experiments of their own. They used the change in specific heat, almost as an afterthought, to explain that some of the heat evolved in the formation of fixed air was absorbed by the blood without raising its temperature. And in a manner very similar to Crawford's they went on to explain that in the circulation when the blood regains the base of fixed air its specific heat is diminished and thus it evolves sensible heat in all parts of the body.[31] It is hard to tell why Lavoisier and Laplace referred to the changing specific heat of the blood, for it is

[30] *Ibid.*, p. 406.
[31] *Ibid.*

clearly an alternative scheme to the one they first offered. In neither system of heat transfer within the body did they provide any experimental evidence; their primary aim had been to demonstrate the similarity between combustion and respiration; the secondary question of what happened to the heat within the body remained an unsolved problem, and was later to become one of the points upon which critics attacked Lavoisier's theory.[32]

In the period immediately following the first publication of the *Mémoire sur la Chaleur* (1783), Lavoisier continued the calorimetric measurements of animal heat and combustion, and also analyzed the alterations in the air brought about by several different types of combustion, as well as by the respiration of animals. A notebook entry for any single day might find Lavoisier at work with the calorimeter measuring the heats of combustion of a candle and of carbon, and that produced in the respiration of a guinea pig.[33] A look at the change in materials that Lavoisier worked with is indicative of the change in direction that his thoughts about respiration were taking. From the original experiments in which respiration had been compared to the combustion of carbon and charcoal a switch occurred in which a candle or wax was substituted for the charcoal.[34] Then we find Lavoisier measuring not only the amount of carbonic acid gas (fixed air) formed but also the amount of water.[35]

The thought behind the experimental changes became apparent when Lavoisier read his "Mémoire sur les altéra-

[32] See Mendelsohn, "Site of Heat Production in the Body," *Proc. Amer. Phil. Soc.*, 105 (1961), 412–420.
[33] See Daumas, *Lavoisier*, for entries from the Lavoisier notebooks; see, for example, p. 51.
[34] By December 1783 Lavoisier was using a candle; *ibid.*, p. 51.
[35] June 1784; *ibid.*, p. 53.

tions qui arrivent à l'air," in February 1785.[36] In this paper
Lavoisier noted that when vital air was respired something
less than the predicted volume was recovered after the
generation of the fixed air.[37] He was forced to conclude
that a part of the vital air is neither converted to fixed air
in the lungs nor restored to its elastic state. Two alternative
interpretations presented themselves to Lavoisier: one, that
the missing portion of the vital air may combine with the
blood; two, that the vital air had combined with a portion
of the inflammable air in the body and formed water in
the lungs.[38] An indication that Lavoisier had not made up
his mind on what products were formed in respiration, or
on where in the body this formation occurred is his citing
his belief that the air extracts something from the lungs
during respiration, and this must be a carbonaceous substance since fixed air was formed by respiration.[39]

Thus by 1785, although Lavoisier was sure that respiration was a slow form of combustion, he was undecided on
a number of questions dealing with how and where heat
was generated within the animal body. He was not sure
whether carbon alone or carbon and hydrogen were involved in the alteration of the vital air. He suggested as one

[36] Lavoisier, "Mémoire sur les Altérations qui Arrivent à l'Air dans
Plusieurs Circonstances où se Trouvent les Hommes Réunis en Société,"
Hist. de la Société Royale de Médecine, 1782–3 [1787], pp. 569–582.

[37] Lavoisier, "Les Altérations à l'Air," p. 574. In July 1784 Lavoisier
had completed a review of the facts related to the formation of fixed air:
"Mémoire sur la Formation de l'Acide, Nommé Air Fixe ou Acide
Crayeux, et que Je Désignerai desormais sous le Nom d'Acide du
Charbon," *Mém. Acad. R. Sci.*, 1781 (1784), pp. 448–467. On the basis
of these earlier findings Lavoisier had predicted that he would find
1498½ cubic inches of vital air, but actually found only 1443⅔, indicating
a deficit of some 54 cubic inches.

[38] Lavoisier, "Les Altérations à l'Air," p. 574. This idea was clearly an
outgrowth of the recent (1783–4) discovery of the composition of water.
See J. R. Partington, *A Short History of Chemistry* (London, 1948),
pp. 140–145.

[39] Lavoisier, "Les Altérations à l'Air," p. 573.

possible alternative that some vital air united with the blood in the lungs. Although he still assumed that the fixed air was formed in the lungs, he had previously suggested that the heat may not all be generated in the respiratory organs.

These problems, however, were not always the ones that caused some people to reject Lavoisier's theory. Writing shortly after the publication of the "Mémoire" of Lavoisier and Laplace, Fabre complained that these two authors were principally concerned with measuring heat in general, not only animal heat. Fabre did not doubt that the combination of pure air with the base of fixed air produces heat, but he could not believe that this phenomenon was applicable to animal heat. As an alternative Fabre described a theory of animal heat based on the irritability of tissues and organs, especially the heart, whose innumerable fibers are in continual collision. A theory of this type, he believed, was beyond the understanding of the chemist.[40] Fabre was clearly reflecting the outlook of those students of nature who insisted that the explanation of the processes of the living organism must of necessity be of a different sort from that provided for the physical world. The revolution being wrought by Crawford, Lavoisier, and their associates seemed, without even directly raising the question, to put an end to the assumed difference between the organic and inorganic realms. The strength of the new move was its basis in experiment rather than philosophical argument.

Although Fabre was not the only physiologist to reject the respiration theory of animal heat,[41] it seems clear, from our modern outlook, that the most successful experimen-

[40] P. Fabre, *Réflexions sur la Chaleur Animale, pour Servir de Supplement à la Seconde Partie des Recherches sur Différens Points de Physiologie* . . . (Paris, 1784), pp. 3–4, 8, 19.

[41] See below for some comments about the adherents of the "nerve theory" of animal heat.

talists were those who pursued answers to the questions raised by Crawford's and Lavoisier's proposals. Crawford himself returned to the study of animal heat with renewed confidence in the efficacy of his theory, and also with an important series of supplemental experiments.

In a paper prepared for the Royal Society in 1781 Crawford was ostensibly dealing with the recently noticed ability of animals to keep themselves at a temperature lower than the environment.[42] While experimenting on this matter, Crawford noticed that after a dog had been immersed in warm water for half an hour a remarkable change was produced in the appearance of the venous blood.[43] Normally the venous blood was a dark red by comparison with the light red of arterial blood. However, after the dog's immersion in the warm medium the venous blood had assumed nearly the hue of the arterial. The implications of this observation, Crawford claimed, confirmed an opinion first suggested to him by "Mr. Wilson of Glasgow":[44]

Admitting that the sensible heat of animals depends upon the separation of absolute heat from the blood by means of its union with the phlogistic principle in the minute vessels, may there not be a certain temperature at which that fluid is no longer capable of combining with phlogiston, and at which it must of course cease to give off heat?[45]

Thus the blood, being less phlogisticated, would maintain a paler hue, since Priestley had demonstrated that the dark color of venous blood was due to the phlogiston with which it became impregnated.

[42] Adair Crawford, "Experiments on the Power that Animals, when Placed in Certain Circumstances, Possess of Producing Cold," *Phil. Trans. Roy. Soc. (London)*, 71 (1781), 479–491.

[43] *Ibid.*, p. 487.

[44] Presumably the Patrick Wilson who is mentioned in the first edition of Crawford's *Animal Heat*.

[45] Crawford, "Power of Producing Cold," p. 488.

With the aid of this process, Crawford believed, the animal could actually cool itself; for an increase in the temperature of the environment causes an increased evaporation in the lungs, which in turn greatly diminishes the sensible heat of the blood in the pulmonary vein. In consequence, this blood will be able to absorb heat from the surrounding vessels and, since this heat will not be liberated in the capillaries, "a positive cold will be produced . . . Thus it appears," wrote Crawford, "that when animals are placed in a warm medium, the same process which formerly supplied them with heat becomes for a time the instrument of producing cold, and probably preserves them from such rapid alterations of temperature as might be fatal to life."[46] Crawford has given a solution (ingenious even if incorrect) to the problem which had puzzled Boerhaave and after him numbers of other physicians and physiologists.[47] Incidentally to his answer Crawford made an observation about the color of blood which was to be of major importance in later metabolic studies.[48]

When Crawford republished his *Experiments and Observations on Animal Heat* in 1788 it did indeed contain

[46] *Ibid.*, pp. 489–490.

[47] See for instance, Blagden, "Experiments in an heated Room."

[48] See Eduard Farber, "The Color of Venous Blood," *Isis*, 45 (1954), 3–9. The experiments on the relation between the state of the blood (color and phlogiston content) and the temperature of the environment were continued by Crawford and reported to the Royal Society in a paper on "Experiments and Observations on the Stability of Heat in Animals" on December 6, 1787, but were not submitted for publication. The manuscript was discovered in the Archives of the Royal Society thanks to a hint from the Librarian N. H. Robinson. *Journal Book of the Royal Society*, 33 (1787–1790), p. 60. The text has been appended to my bibliobiographical study of Adair Crawford. The observation made by J. R. Mayer half a century later that venous blood in the tropics was redder than that found in a man's veins in a temperate climate led him to propose a theory of conservation of energy. See Stacey B. Day, "A Controversial Fruit of a Voyage to Java," *Surgery, Gynecology and Obstetrics* (March 1961), pp. 381–385.

the "very large additions" which the title page claimed for it.[49] The most significant change, however, was one of technique rather than theory. Crawford had developed a calorimeter which he used to determine the amount of heat generated by combustion and respiration. Although Crawford indicated that he had read Lavoisier and Laplace's "Mémoire sur la Chaleur," his own apparatus was different in principle from the one described in the "Mémoire." As a matter of fact, Crawford was critical of the ice calorimeter and did not recognize the "correction" that Lavoisier and Laplace had been forced to make in order to find a correspondence between the heats of combustion and of respiration. In Crawford's literal interpretation of Lavoisier's experiment it appeared that when equal portions of air are altered by a guinea pig and by combustion of charcoal the quantity of heat produced by the guinea pig is to that produced by the charcoal as 13 to 10.3. The reason for this lack of correspondence was quite apparent to Crawford; the determination of the amount of air altered by respiration had been carried out in a vessel at 60°F, while the heat generated by the animal was measured when the ambient temperature was 32–33°F.[50] Crawford had already demonstrated in his 1781 paper that animals in colder atmospheres phlogisticate more air.[51] Thus, an inaccuracy was caused by the very apparatus utilized by Lavoisier and Laplace.

The calorimeter that Crawford built was probably designed to avoid the inaccuracies he had criticized in

[49] The second edition was significantly larger. The 1779 edition had some 120 pages, the 1788 edition contained 491 pages plus plates. The second edition is dedicated to the Irish chemist Richard Kirwan.

[50] Crawford, *Animal Heat* (1788), pp. 147n, 332, 333, 375.

[51] Crawford, "Power of Producing Cold," pp. 489–491. This point is underscored in the additional Crawford experiments reported in 1787. See footnote 45 above.

Lavoisier and Laplace's apparatus. In place of ice Crawford used water, and thus determined the quantity of heat produced by measuring the changes in the temperature of the water in the vessel surrounding the respiration-combustion chamber.[52] He was also able to analyze the alterations of the air that took place within the calorimeter and was therefore assured that the heat production was directly correlated with the evolution of fixed air, both being measured when the animal was in an environment at the same temperature.[53]

Crawford, like Lavoisier, seemed inclined to believe that inflammable air (hydrogen) as well as carbon, or the base of fixed air, was involved in the alteration of pure air in the lungs.[54] On the basis of earlier studies by Henry Cavendish, Crawford concluded that "it is possible, that a portion of the pure and inflammable air, which meet in the lungs, may undergo that peculiar mode of combination by which water is produced."[55] At the suggestion of Priestley, Crawford attempted to determine the heat of combustion of inflammable air by igniting it and pure air within a brass chamber so arranged that heat increase could be measured.[56] Finding what he believed to be a partial conversion of pure air to water vapor in both respiration and the burning of a wax taper, Crawford began comparing the heats produced by the two processes. Charcoal, on the other hand, produced no aqueous vapor when burned. The amounts of heat generated by these three sources in altering the same

[52] Crawford, *Animal Heat* (1788), p. 319.

[53] It should be pointed out that Crawford also made determinations of the respiratory alteration of the air in a separate apparatus consisting of two bell jars (*ibid.*, p. 324).

[54] *Ibid.*, pp. 148, 154. Priestley is cited as the source of this suggestion.

[55] *Ibid.*, p. 154.

[56] *Ibid.*, p. 254. It should be noted that in this second edition Crawford has adopted much of the French chemical nomenclature.

quantity of pure air were: wax, 21; charcoal, 19.3; guinea pig, 17.3. The quantities of heat were judged to be close enough, by Crawford, to allow the conclusion that they could be considered equal. The differences were put down to the fact that more heat is generated in the conversion into water than into fixed air, and that the guinea pig lost some heat through "insensible perspiration." Crawford tended, however, to equate respiration with the combustion of wax, especially because of the water produced in both cases.[57]

Crawford joined Lavoisier in doubting that any dephlogisticated, or pure, air is absorbed by the blood in its pulmonary passage, but he did not accept Lavoisier's account of the liberation of heat in the lungs. The blood in passing through the lungs, Crawford claimed, undergoes a change similar to that which occurs when solid bodies are melted; the capacities for heat are increased in both cases. As solids melt they absorb a quantity of heat; similarly, when venous blood is changed to arterial blood an absorption of heat takes place. Crawford had not had to rely upon some "humidity" of the lungs to absorb the heat being released from the pure air. Further, he stated, he did not believe, as had Lavoisier, that elementary fire is chemically combined with bodies or that the heat is disengaged from the air in respiration by a double elective attraction, for before this is possible it must be shown that heat is a substance, and this has not yet been demonstrated.[58] Heat, he stressed, is evolved from the air in consequence of an alteration in the capacity of the air; this is the case for both

[57] *Ibid.*, pp. 351–353, 359–360.

[58] *Ibid.*, pp. 155n, 360, 363. Crawford has taken a much more conservative line in this edition regarding heat as a substance. He pointed out, however, that the "experimental facts" he presented and the consequences of them were in no way dependent upon the various hypotheses on the nature of heat. He still tended to favor the notion of heat as a substance (pp. 435–436).

animal heat and that generated by the inflammation of combustible bodies.[59]

What emerged when Crawford was through was a chemical and physical theory of animal heat, based upon a wide range of experimentation in which similar results were claimed for living and nonliving systems. In point of fact, the distinction between these two falls away in the pages of Crawford's book. The blood, once the reservoir of life, is treated as any other fluid might be, undergoing changes in capacity for heat, absorbing and releasing inflammable airs. No major attack has been launched against "the vital principle"; instead, it has just been assumed that the body behaves according to chemical and physical laws.[60]

This outlook was certainly at the base of Lavoisier's work and in the few years prior to his untimely death he sought to extend this chemical approach to digestion and transpiration, two processes which together with respiration governed "la machine animale."[61] Working with Lavoisier, on what amounts to a series of fundamental studies on animal metabolism, was the physiologist Armand Séguin.[62] In a review article dealing with respiration and

[59] *Ibid.*, p. 380.

[60] I would not split philosophical hairs as G. J. Goodfield does in her study, *The Growth of Scientific Physiology. Physiological Method and the Mechanist-Vitalist Controversy, Illustrated by the Problems of Respiration and Animal Heat* (London: Hutchinson, 1960), p. 63. In method and operation Crawford has assumed that the physiological system is indistinguishable from a physical-chemical system. His failure to directly enter the philosophical arguments over vitalism should not be misconstrued. Most working biologists remained outside of this discourse.

[61] Armand Séguin and Antoine Lavoisier, "Premier Mémoire sur la Respiration des Animaux," *Mém. Acad. R. Sci.*, 1789 [1793], p. 580.

[62] Séguin's collaboration with Lavoisier began sometime prior to July 30, 1790, for on that day, while reading a report to the Société Royale de Médecine, Séguin announced that he and Lavoisier would shortly publish a paper on respiration. Denis Duveen and Herbert Klickstein, *A Bibliography of the Work of Antoine Laurent Lavoisier, 1743–*

animal heat, written just prior to his collaborative works with Lavoisier, Séguin gave an indication of the direction their efforts would take in solving some of the problems raised by Crawford's and Lavoisier's theories.[63] Séguin briefly outlined the history of Crawford's and Lavoisier's work and attempted to strengthen some of their conjectures. Whereas Lavoisier had only hinted in his 1785 memoir that vital air combined with hydrogen in the lungs, Séguin marshaled the available evidence to show that this is the case. He cited the work of Cigna, Priestley, and Hamilton as well as experiments of his own showing that the change of the color of blood from arterial to venous can be effected by the introduction of hydrogen.[64]

It could be objected, however, that it is hard to conceive how hydrogen and carbon are combined with vital air in lungs. After all, it was pointed out, it takes the application of a burning body to inflame hydrogen and a heat of nearly $150°R$ is necessary to inflame carbon in the air.[65] In turning to answer this problem Séguin was dealing with the single most difficult objection to the conception of respiration as a slow form of combustion. The heats normally generated by the combustion of hydrogen and carbon came to be counted on in the computation of the sources of animal heat, yet it was quite clear that the explosion produced by the inflammation of hydrogen could not be tolerated in

1794 (London: William Dawson and Sons, and E. Weil, 1954), p. 83. The *Procès-Verbaux* of the Académie des Sciences, 13 November 1790, report that Lavoisier and Séguin have formulated a plan for an extended study of almost all parts of the animal economy. Daumas, *Lavoisier*, p. 64.

[63] Séguin, "Observations Générales sur la Respiration et sur la Chaleur Animale," *Observations sur la Physique*, 37 (1790), 467–472. This paper was read to the Académie on 22 May 1790. Daumas, *Lavoisier*, p. 64.

[64] Séguin, "Observations Générales," pp. 468–469.

[65] *Ibid.*, pp. 469–470.

the lungs, nor could the excessive heats usually found to accompany the burning of carbon. Séguin, however, was clearly not of a mind to attribute these phenomena to a special animal power, and instead came up with an alternative "chemical" solution.

What had been a mere suggestion in Lavoisier's 1785 memoir, that the vital air extracted some carbonaceous substance from the lungs during respiration,[66] became for Séguin a chemical combination of hydrogen with carbon (*hydrogène carboné*) which was released by the blood into the lungs. Because the hydrogen is not freed in the lungs in a gaseous state, it is not prevented by its attraction for caloric (or heat) from uniting with vital air at ordinary temperatures; and similarly, explained Séguin, because the carbon is held in a very divided state by its combination with hydrogen it too is able to combine readily with oxygen at body temperature.[67] Thus Séguin proposed the means whereby "combustion" can occur in the lungs without a superabundance of heat being generated; but it is a proposal with little or no experimental basis.

Unlike Lavoisier, who made scant use of Crawford's findings about the difference in capacity for heat of arterial and venous blood, Séguin adopted the idea almost in its entirety. He simply substituted the new terminology, suggesting that in the lungs the attraction of "hydrogène carboné pour l'oxigène" is greater than the attraction between oxygen and caloric or the attraction of carbonated hydrogen for the blood. The vital air (oxygen + caloric) is decomposed during inspiration and the caloric which is released unites with the blood which has had its capacity

---

[66] Lavoisier, "Les Altérations à l'Air," p. 573.

[67] Séguin, "Observations Générales," pp. 469–470. Bertholet and Priestley are cited as having demonstrated that hydrogen and oxygen unite at ordinary temperatures.

for heat augmented by the loss of some carbonated hydrogen.[68] The reverse process occurs in the systemic circulation. Thus, at the very time that Crawford was adopting the language of the French school, the latter were incorporating his suggestions of how heat is transferred from the air into the animal body.

Lavoisier and Séguin obviously intended a joint exploration of the three regulating principles which they believed governed "la machine animale." They succeeded in studying respiration and transpiration, and although they began their research on digestion[69] it never came to fruition in a separate published work. The four joint papers which were presented detail a number of remarkable experiments, but indicate a slow development of Lavoisier's earlier theories rather than any radical departures. In appraising past work they noted that in the "Mémoire sur la Chaleur" Lavoisier and Laplace had shown, without recognizing its importance, that animals in a given time release a greater quantity of caloric than would be expected from the amount of carbonic acid formed in an equal time by respiration.[70] Séguin and Lavoisier were convinced by the time they wrote in 1790 that the excess heat was due to the combustion of hydrogen.

An experimental finding of some interest was one that tended to show a difference between combustion and respiration. It had long been known that combustion occurs more rapidly when the air in which it takes place is purer. That is, in a given time more carbon will burn in vital air

---

[68] *Ibid.*, p. 471.

[69] The *Procès-Verbaux* of the Académie, 10 June 1791, note that Lavoisier and Séguin have commenced the study of digestion. Daumas, *Lavoisier*, p. 65.

[70] Séguin and Lavoisier, "Premier Mémoire sur la Respiration," p. 569. When Lavoisier and Laplace originally wrote they attributed the excess heat to experimental error. Crawford had also noticed the discrepancy. See above.

than in atmospheric air. Although they had thought the same would be true in the case of respiration, experiment destroyed the notion, for they found that, regardless of whether an animal respired in vital air or in vital air mixed with a portion of azote (nitrogen), the amount of vital air consumed remained constant.[71] When they reflected on this point in their "Second Mémoire" they placed a great importance on the fact that the amount of vital air consumed in respiration was always the same, regardless of the source of the air.[72] The authors expressed great praise for nature and its ability to maintain an equilibrium in which only enough hydrocarbonous fluid was burned to provide a constant body temperature, in this being unlike a fire which would burn faster as more vital air was available.[73]

In these studies the *hydrogène-carboné*, introduced by Séguin in his paper of 1790, became a humor which was continually separated from the blood in the bronchial tubes and filtered through the membrane of the lungs. Within the lungs, the humor, composed primarily of hydrogen and carbon, burns and decomposes the vital air.[74] But this is not the complete story, for Lavoisier and Séguin were not quite willing to assert that the lungs were the site of carbonic acid formation. It was still possible that some oxygen was absorbed by the blood and combined with carbon during the circulation. They also were willing to consider the possibility that some of the carbonic acid was formed during digestion and secreted into the blood along with the chyle.[75] In retrospect, the most interesting sugges-

---

[71] *Ibid.*, p. 573.
[72] Lavoisier and Séguin, "Second Mémoire sur la Respiration," *Ann. de Chimie*, 91 (1814), 333. Originally read to the Académie on 9 April 1791, but not published during Lavoisier's lifetime.
[73] *Ibid.*
[74] Séguin and Lavoisier, "Premier Mémoire sur la Transpiration des Animaux," *Mém. Acad. R. Sci.*, 1790 [1797], p. 606.
[75] Séguin and Lavoisier, "Premier Mémoire sur la Respiration," p. 583.

tion regarding the site of carbonic acid production arose out of the studies on transpiration. Lavoisier and Séguin had found that water and carbonic acid were formed at the surface of the skin of the whole human body, from which they concluded that animals do not respire in the lungs alone.[76] The skin, like the lungs, also had seeping on to it a viscous humor which was hydrocarbonous in nature.[77] What was never explicitly stated, if thought of at all, was that the surface of the skin as the site of carbonic acid and water production was in actuality a site of combustion, and consequently some of the animal heat should be generated there.

To the historian, the achievement of Lavoisier, and to a lesser extent of Crawford, in founding the oxidation theory of combustion and animal heat is the single most important innovation in the long history of attempts to account for the warmth of animals. It was only after the experimental successes of Lavoisier that one could look back and "rediscover" the combustion theory of Mayow and recognize the conceptual similarity between it and the oxygen theory of Lavoisier.[78] But the specific differences are marked

[76] Séguin and Lavoisier, "Second Mémoire sur la Transpiration des Animaux," in Lavoisier, *Traité Elémentaire de Chimie* (3rd ed.; Paris, 1801), II, 236–237. The *Procès-Verbaux* of the Académie indicate that this paper was read on 22 February 1792: "M. Séguin a lu un *Mémoire sur la transpiration*. M. Lavoisier a annoncé à cette occasion que ce travail commencé d'abord en commun avec lui avait été suivi ensuite par M. Séguin seul et que les expériences de ce mémoire lui appartenaient entièrement." Daumas, *Lavoisier*, p. 65.

[77] Séguin and Lavoisier, "Second Mémoire sur la Transpiration," p. 236.

[78] Two authors who attempted a revival of Mayow shortly after Lavoisier's success were Thomas Beddoes, *Chemical Experiments and Opinions Extracted from a Work Published in the Last Century* (Oxford, 1790), and Grant David Yeats, *Observations on the Claims of the Moderns, to some Discoveries in Chemistry and Physiology* (London, 1798).

enough to need no further comment and to leave little room for the belief that Mayow was the unheralded precursor of Lavoisier's discoveries. It is only after the oxygen theory of combustion had been well founded that one could sympathetically view Mayow's attempts to link combustion and respiration.

Unlike the earlier attempts of Boyle, Hooke, and Mayow to link combustion and respiration, the new efforts of Crawford and Lavoisier were marked by the criterion which identifies a revolutionary scientific advance: people paid attention to the new idea. They attacked it, they gathered evidence to support, but they did not ignore it.

Of equal importance to the attention paid the new theory of animal heat was the general acceptance of the new framework within which this area of physiology, at least, had to be discussed. A physical and an organic process were identified as being nothing more than variations of each other. The terms of discussion of this "vital" phenomenon became identical to those utilized in the physics and chemistry of combustion. To be successful in further studies of animal heat the physiologist had to be able to manipulate the newest theories and concepts of the physical sciences. But the other effect of this change was to arouse a new interest in biological problems on the part of men whose major training and preparation were in physics and chemistry. Indeed, the next pages demonstrate the extent to which physical scientists became responsible for the development of new ideas, and the execution of new experiments, in physiology.

· V I I ·

Physics, Chemistry, and Physiology

Developments in the theory of animal heat in the years immediately following Lavoisier's and Crawford's proposals dealt primarily with problems raised by the writings of these two men. There was an extended discussion of the adequacy of the heats of combustion of hydrogen and carbon to account for the production of all the heat generated by the animal.[1] A controversy arose over the question of where in the animal's body the combustion took place. Lavoisier's followers did not always add the note of speculation that was found in his own discussions of the site of formation of the products of respiration (or combustion within the body). Pierre Adet, in a response to Priestley's attack on the new chemistry, pointed to the carbonic acid produced in the lungs during respiration as the result of a combination between oxygen and carbon secreted from the blood. So that no doubts might linger, Adet noted that in

[1] In 1821, for instance, Lavoisier's own Académie Royale des Sciences proposed that its Prix de Physique for 1823 be awarded for the determination "par des expériences précises quelles sont les causes, soit chimiques, soit physiologiques, de la chaleur animale." *Ann. Chim. Phys.*, 19 (1821), 327. The prize was awarded to Despretz [*Ann. Chim. Phys.*, 23 (1823), 193] for his "Recherches Expérimentales sur les Causes de la Chaleur Animale," *Ann. Chim. Phys.*, 26 (1824), 337–364. Interestingly enough, his results indicated that combustion (respiration) could account for only 70–90 percent of the total heat emitted by the animal (*ibid.*, p. 361).

several diseases of the respiratory organs patients expectorated carbon in pure form along with the mucus of the lungs. Adet claimed that in an illness caused by nervous fatigue, which he himself experienced, he had found his spittle to be black and upon searching for the cause soon found that it was carbon that had caused the color.[2] A similar explanation was proposed, by a Dr. Bree, for the dark-colored substance which is sometimes found in the mucus of the bronchial glands. The good doctor believed that asthmatic patients were not sufficiently able to carry off the carbon secreted into the lungs by the formation of carbonic acid.[3]

A question just as perplexing as where in the body combustion took place was at what temperatures the products of combustion, carbonic acid and water, actually could be formed. Although Séguin had argued that these combustions could occur at ordinary body temperature there was hardly unanimity among the investigators of the problem. Hugh Moises, in his *Treatise on the Blood*, was quite emphatic in his statement:

> We know of no experiment which might authorize us to suppose that carbon can unite with oxygene in a temperature of 97°–99°, or that hydrogene and oxygene airs combine and form water in so low a temperature.[4]

If respiration was to be considered a form of combustion this was a difficult challenge to answer. Daniel Ellis was cognizant of the type of objection raised by Moises, but he

[2] P. A. Adet, *Réponse aux Réflexions sur la Doctrine du Phlogistique & de la Décomposition de l'Eau* (Philadelphia, 1797), p. 77.

[3] Cited by John Bostock, *An Essay on Respiration*, Parts I and II (Liverpool, London, 1804), p. 225.

[4] Hugh Moises, *A Treatise on the Blood, or, General Arrangement of Many Important Facts, Relative to the Vital Fluid. With Some Cursory Observations on the Theory of Animal Heat. Interspersed with Pathological and Physiological Remarks from the Inductions of Modern Chemistry* (London, n. d. [1794]), p. 192.

pointed out that everywhere in nature processes were to be observed which produce carbonic acid at low temperatures. This gas he found produced by the living functions of animals and plants as well as by the decomposition of animal and vegetable matter. From these observations he was led to infer that the carbonic acid, whether produced in the lungs or in the blood, did not arise from a process analagous to combustion.[5] Doubts of this sort were not uncommon among the physiologists of the late eighteenth and early nineteenth centuries.

One group of physiologists became interested in the possible role of the brain and nervous system in the generation of animal heat. Their experiments with nerve sections and brain damage showed a clear relation between the nervous system and metabolic functions; but it was not until well into the nineteenth century that the regulatory activities of the nerves was recognized. In the meantime these experimentalists, often suspicious of the physical and chemical analyses applied to the organism and more anxious to establish separate "biological" laws to explain living things, cast doubt on the other work. Bernard Brodie gave voice to what was not a unique challenge:

Where so many and such various chemical processes are going on, as in the living body, are we justified in selecting any one of these for the purpose of explaining the production of heat?[6]

[5] Daniel Ellis, *An Inquiry into the Changes Induced on Atmospheric Air, by the Germination of Seeds, the Vegetation of Plants, and the Respiration of Animals* (Edinburgh, 1807), p. 124.

[6] "Further Experiments and Observations on the Influence of the Brain on the Generation of Animal Heat," *Phil. Trans. Roy. Soc. (London)*, 102 (1812), 391. See also Brodie, "The Croonian Lecture, on Some Physiological Researches, respecting the Influence of the Brain on the Action of the Heart, and on the Generation of animal Heat," *ibid.*, 101 (1811), 36–48; Enoch Hale, *Experiments on the Production of Animal Heat by Respiration. An Inaugural Dissertation, Read and Defended at*

Brodie's question, however, was not of the adequacy of a given chemical process to account for the animal heat, but rather of the ability of any chemical process to provide acceptable explanation of a biological phenomenon. He proposed no test of the chemical theories in physiology and conducted no experiments to ascertain their applicability. Like many other life-oriented biologists, he developed a life-oriented theory and with it was led to a type of experimentation on living organisms quite different from that conducted by the chemically oriented physiologists of the very same period. The expectation was that special biological explanation would be needed for organic processes and in consequence the expectation was fulfilled by the very nature of the experiments conducted.

One of the earliest friendly critics of the Lavoisier-Crawford theory[7] of animal heat was Christoph Girtanner, the man responsible for introducing Lavoisier's chemical doctrines into Germany. Girtanner, writing in 1790, in a paper in which he attempted to show that oxygen is the principle of irritability (which in turn is the principle of life) took issue with several aspects of the combustion theory of animal heat. First, he objected that he knew of no experiment showing that carbon and oxygen can be united at temperatures found in the body, nor was there any experiment showing hydrogen and oxygen to combine at 30 degrees.

---

the *Public Examination, Before the Rev. President and the Medical Professors of Harvard University,* August 20, 1813 (Boston, 1813); Everard Home, "The Croonian Lecture On the Changes the Blood Undergoes in the Act of Coagulation," *Phil. Trans. Roy. Soc. (London),* 108 (1818), 172–184; Julian Jean César Legallois, "Deuxième Mémoire sur la Chaleur Animale," *Ann. Chim. Phys.,* 4 (1817), 5–23, 113–127; Charles Chossat, *Mémoire sur l'Influence de Système Nerveux sur la Chaleur Animale* (Paris, 1820).

[7] I have linked the two theories together, since this is generally what has been done by most authors post 1790.

He was not impressed by Séguin's attempts to answer these objections by proposing that the carbon was in a very divided state owing to its combination with hydrogen. Girtanner judged this response to be hypothetical and unconvincing.[8] Instead he suggested that part of the respired oxygen is absorbed by the blood itself and consequently increases the blood's capacity for heat.[9] During the circulation the blood loses its oxygen (which for Girtanner as for Lavoisier is oxygen plus caloric) and takes up a carbonated hydrogen gas through a double affinity. The caloric which has been distributed with the oxygen is released in the system, thus providing the animal heat.[10]

But, still sticking fairly closely to Lavoisier's scheme, Girtanner believed some of the vital air was decomposed in the lungs, thus forming carbonic acid and also liberating caloric. This free caloric, claimed Girtanner, produces the temperature necessary for the formation of water by the combination of oxygen gas and hydrogen gas. Girtanner has thus constructed a system which contains characteristics of both Crawford's and Lavoisier's proposals, but in trying to avoid the difficulties raised by his predecessors he stumbled into one of his own, that of having to deal with the great heat necessary for the combustion of hydrogen. For this he proposed no explanation.

[8] Christoph Girtanner, "Mémoires sur l'Irritabilité, Considerée comme Principe de Vie dans la Nature Organisée, Second Mémoire," *Observations sur la Physique*, 37 (1790), 140.

[9] This variation on the Crawford theme (actually just the reverse of what Crawford had to say if the change in nomenclature is followed) is claimed by Girtanner to be a new truth which he generalized to say that oxygen raises the capacity for heat of any substance with which it combines. He hoped to prove this in a later memoir, *ibid.*, p. 141, 141n.

[10] *Ibid.*, p. 147. In a footnote Girtanner indicated that he hoped to present a full memoir on animal heat. I do not know of the existence of such a paper.

The experiments that Girtanner carried out were primarily designed to demonstrate that oxygen was absorbed by venous blood and was the factor responsible for its transformation into arterial blood.[11] The two elements in Girtanner's proposal which were important for the theory of animal heat, that oxygen is absorbed by the blood and that the heat is released from the blood during circulation, were not originated by him; one had been conjectured by Lavoisier and the other contained in Crawford's theory. However, bringing the two proposals together in the manner he did foreshadowed an important stream of new research into the site of combustion in the animal body. Even Girtanner's crude beginnings indicated that much could be learned from the analysis of the gases found in the blood.

Recognizing the limited nature of Girtanner's solution to the problem, Joseph Louis Lagrange, the mathematician, proposed that the blood in passing through the lungs absorbed all the respired oxygen.[12] In a memoir by Jean Hassenfratz, Lagrange is quoted as having objected to the inadequacy of Lavoisier's proposal which would have local-

[11] *Ibid.*, p. 145f.

[12] Jean Hassenfratz, "Mémoire sur la Combinaison de l'Oxigène avec le Carbone et l'Hydrogène du Sang, sur la Dissolution de l'Oxigène dans le Sang, et sur la Manière dont le Calorique se Dégage," *Ann. Chimie*, 9 (1791), 266f. Just why Lagrange, like Laplace a mathematician, became involved in this physiological controversy is hard to tell. Icilio Guareschi, "Notizie storiche intorno a Luigi Lagrange," *Mem. R. Accad. Torino*, 64 (1914), 1–13, has been unable to locate any writings of Lagrange dealing with the question of animal heat or respiration, claiming that all we have comes through the article by Jean Hassenfratz whom he identifies as an assistant in Lavoisier's laboratory and a friend of Lagrange. George Sarton, "Lagrange's Personality (1736–1813)," *Proc. Amer. Phil. Soc.*, 88 (1944), 457–496, warned against uncritical biographers who attribute too much scientific curiosity to Lagrange. Sarton suggested that Lagrange's fears concerning his own health had caused him, early in life, to obtain some knowledge of physiology and medicine, and that this knowledge was gradually improved.

ized the release of all the heat in the lungs, and only later distributed it throughout the animal system.[13] If this were the case, Lagrange complained, the temperature of the lungs would be elevated to such a degree that there would be constant fear of their destruction. Furthermore, the lung temperature would be so different from that of other parts of the body that it would be impossible that no one had as yet noticed it. Lagrange believed it much more probable that the animal heat was released not only in the lungs but in all the parts of the body reached by the circulating blood. This would be achieved if all the oxygen were dissolved in the blood during respiration and thus carried into the arteries and veins. In the extremities of the body the oxygen would slowly leave the state of dissolution and combine with the carbon and hydrogen of the blood, thus liberating caloric and also forming the carbonic acid and water which are released in the lungs.[14] Lagrange thus added a new element to the theory of animal heat in postulating that the total combustion or oxidation takes place in the blood in circulation rather than in the lungs.

Lagrange was critical of a view which he attributed to Laplace, but which is also found in Crawford, that suggested that only a part of the heat was released in the lungs in a sensible state, the rest remaining combined in a latent form, being released only during the circulation.[15] This

[13] Hassenfratz, "Combinaison de l'Oxigène," p. 266. Berthelot, "Remarques sur un Point Historique Relatif à la Chaleur Animale," *Comptes Rend. Acad. Sci.*, 77 (1873), 1065, noted Lagrange's objection and suggested that actually the amount of heat which would be liberated were combustion to occur in the lungs would not be enough to raise the temperature of the lungs more than a fraction of a degree. Lagrange's basis for objection was wrong, "mais ce n'est pas la seule fois dans l'histoire des sciences qu'un argument sans valeur est devenu l'origine de découvertes importantes."

[14] Hassenfratz, "Combinaison de l'Oxigène," pp. 266, 272.

[15] *Ibid.*, p. 267.

explanation, Lagrange complained, does not rest on fact, but only on the distant analogy of the absorption of caloric by bodies as they pass from a solid to a liquid state, or from a liquid to a gaseous state.[16] This is clearly not the case with blood passing through the lungs. Lagrange, however, was quite willing to believe, along with Girtanner, that all oxidized substances have a greater comparative heat than pure substances, and that arterial blood as compared to venous blood can be considered oxidized. He further explained that the oxygen dissolved in the arterial blood still has all its caloric in it, which is ultimately released in the circulation as the oxygen enters combination. On the other hand, the venous blood, in which the oxygen has already combined, has lost some caloric. In this manner Lagrange explained Crawford's findings about the difference in capacity for heat of blood.[17]

As one reads it, Lagrange is not arguing about a living system as such. Nor is he really discussing physiology, interpreted as the study of the functioning of living things. Lagrange has accepted the new framework of discourse and addresses his remarks to what is in essence a physiochemical system. He is talking about the interaction of fluids, chemical elements, and heat; it is incidental that they happen to be in an animal's body.

The questions which Lagrange's hypothesis raised were those to which several generations of physiologists and chemists devoted their efforts. They attempted to determine experimentally whether there was oxygen in the arterial

[16] Cf. Crawford above.
[17] Hassenfratz, "Combinaison de l'Oxigène," p. 273f. Hassenfratz contributed to this paper the results of a series of experiments tending to show the verity of Lagrange's hypothesis by demonstrating the existence of oxygen in the arterial blood. He thus initiated the type of blood-gas analysis which was to occupy many chemists and physiologists.

blood and carbonic acid in the venous blood. The controversy over the site of combustion in the body had been
narrowed down to a question of the presence or absence
of gases in the blood. The early studies of this limited
problem suggested solutions based on the ability of the
different gases to cause the color changes observed in
blood. Both Girtanner and Hassenfratz had attempted to
show that venous blood became bright red upon the absorption of oxygen, and thereby to demonstrate their contention that the blood took up oxygen in the lungs. Another
line of experimentation attempted to recover gases from
the blood. Hugh Moises, for instance, confined arterial
blood in bottles with gases containing no oxygen and
claimed to have found oxygen in the residue.[18] In a similar
series of experiments Everard Home reported the separation of carbonic acid from venous blood confined in a
receiver from which the air had been withdrawn.[19] The
work on absorption of gases by liquids carried out by
William Henry and John Dalton probably served to inspire
other attempts to show that the blood took up the respiratory gases in the lungs. Humphry Davy heated arterial
blood to 108°F and collected the liberated gas. Upon analysis he found carbonic acid and "phosoxygen" (oxygen +
light), which proved to him that "phosoxygen" had been
absorbed by the arterial blood.[20] Allen and Pepys, in a
group of carefully executed experiments, made use of a

[18] Moises, *On the Blood*, pp. 207–208.

[19] Home, "Changes of Blood in Coagulation," p. 181.

[20] Humphry Davy, "Essay on Heat, Light, and the Combinations of
Light, with a New Theory of Respiration," (1799), *Collected Works*,
ed. John Davy, Vol. II (1839), p. 78. Davy's new theory of respiration
can be succinctly summed up in the following quotation (p. 79): "Respiration, then, is a chemical process, the combination of phosoxygen
with venous blood in the lungs, and the liberation of carbonic acid and
aqueous gas from it." The animal heat, he believed, arose from the chemical combinations and decompositions taking place in the body.

gasometer (respirometer) to demonstrate an exchange in the blood between azote (nitrogen) and oxygen.[21] They also challenged the previously accepted proposal that water was formed by the combination of a part of the respired oxygen with hydrogen. The quantity of carbonic acid emitted was found to be equal volume for volume to the oxygen consumed "and therefore," they said, "there is no reason to conjecture that any water is formed by a union of oxygen and hydrogen in the lungs."[22]

Although the piling up of evidence in favor of Lagrange's hypothesis should have given it great weight, the proposal probably had as many critics as supporters during the first two decades of the nineteenth century. A theory as well entrenched, and as prominently supported, as Lavoisier's and Crawford's theory of combustion in the lungs did not give way easily. While willing to recognize the fact that gases when in contact with some fluids could become diffused through the fluids, Daniel Ellis rejected the opinion that this was true in the case of oxygen and blood. Ellis contended that the attraction of oxygen by the blood through the coats of the vessels was inconsistent with the laws of chemical attraction and affinity. He denied that there existed any proof which showed air to be present in healthy blood, disputed Davy's conclusions to this effect, and claimed that they were at variance with the established laws of the animal system. After an extended argument Ellis concluded that the animal heat is produced by the combustion of carbon in the lungs and the reason that the lungs are not overheated is, as Crawford suggested, the change in capacity of arterial blood which permits the

21 William Allen and William H. Pepys, "On Respiration," *Phil. Trans. Roy. Soc. (London)*, 99 (1809), 404–429.

22 Allen and Pepys, "On the Changes Produced in Atmospheric Air, and Oxygen Gas, by Respiration," *Phil. Trans. Roy. Soc. (London)*, 98 (1808), 279.

absorption of the excess heat and its distribution to the body in a latent form. Thus, according to Ellis, the controversy can be resolved; the decomposition of air (combustion) occurs in the bronchial cells of the lungs.[23]

Ellis was not alone in his feeling that the new evidence was not sufficient to overthrow the older theories. In the course of an extensive review of research developments in respiration and animal heat, John Dalton, who is best known for his atomic theory of matter, claimed that to his knowledge there were as yet no decisive experiments on the question of whether oxygen is actually absorbed by the blood, with the production of carbonic acid taking place in the process of circulation. He rightly pointed out that many of the seemingly contradictory results could be interpreted so as to be in accord with the theory of Lavoisier and Crawford. Dalton was still inclined to accept the main principles of Crawford's theory: "indeed his results . . . are so plausible, and his whole theory so beautiful, that one would feel a regret in having to question the accuracy of his principles."[24]

During the 1830's a high point was reached in the efforts to resolve the question of whether the carbonic acid was produced in the lungs or in the blood; the old experiments were repeated; better pneumatic apparatus was utilized; contradictory results were harmonized.[25] These efforts culminated in the carefully executed researches of Gustav Magnus. Using several different types of experiment he was able to recover carbonic acid from venous blood; he

[23] Ellis, *Changes Induced on Air*, pp. 124, 142, 234, 235.

[24] John Dalton, "On Respiration and Animal Heat," *Mem. Manchester Lit. Phil. Soc.* [2], 2 (1813), 36, 38–39.

[25] The work of John Davy, younger brother of Humphry Davy, is representative of the inconclusive type of research carried out in the early nineteenth century, which finally turned productive in the 1830's when the question of animal heat was limited to the finding of gases in

subjected the blood to atmospheres of other gases and also to very low pressures achieved with a vacuum pump.[26] Magnus then argued that, if the carbonic acid was already formed in the venous blood, its release from the blood into the lungs should follow the same laws that govern the release of any gas dissolved in a liquid and an equal volume of a new gas should be absorbed by the liquid. A confirmation of this suggestion was found in Magnus's demonstration that arterial blood contained oxygen. He recognized, however, that according to the laws of absorption not all the carbonic acid would be released in the lungs and that some of it must be contained in arterial blood. The results that he ultimately was able to achieve showed that all blood contained oxygen, carbonic acid, and nitrogen. He found that the arterial blood was richer in oxygen than the venous blood, and that the venous blood had a higher proportion of carbonic acid than the arterial blood. It became apparent to Magnus that the carbonic acid was not being produced in the lungs, but rather was generated in the blood during circulation. He suggested that as the oxygen-rich blood passed through the capillaries an oxidation of carbon occurred, producing carbonic acid.[27]

the blood. The following papers by Davy present a view of the changing nature of research on animal heat: "An Account of Some Experiments on Animal Heat," *Phil. Trans. Roy. Soc. (London)*, 104 (1814), 590–603; "Observations on Air Found in the Pleura, in a Case of Pneumatothorax; with Experiments on the Absorption of Different kinds of Air Introduced into the Pleura," *Phil. Trans. Roy. Soc. (London)*, 113 (1823), 496–516; "Observations on the Temperature of Man and Other Animals," *Edinburgh Phil. Journ.*, 13 (1825), 300–311, 14 (1826), 38–46; "An Account of Some Experiments on the Blood in Connexion with the Theory of Respiration," *Phil. Trans. Roy. Soc. (London)*, 128 (1838), 283–299.

[26] Gustav Magnus, "Ueber die im Blute enthaltenen Gase, Sauerstoff, Stickstoff und Kohlensäure," *Ann. Phys. Chem.*, 40 (1837), 583–606.

[27] *Ibid.*, pp. 589, 590, 600–602.

Magnus's work, which received rapid confirmation,[28] answered only the limited question whether carbonic acid was formed in the lungs or in the blood. But since it was widely assumed that the animal heat was due, in large measure, to this combination,[29] the site of body-heat production was conclusively shifted from the lungs to the capillaries. Lagrange's hypothesis had received experimental verification, and Liebig could claim with confidence that "the heat evolved in the process of combustion, to which the food is subjected in the body, is amply sufficient to explain the constant temperature of the body."[30]

It is worth analyzing the changing nature of the questions that had to be answered as the theory of animal heat developed, for these questions are definitely indicative of a much more general change that was occurring in all of physiology. Magnus, for instance, did not have to experiment upon or theorize about the whole organism or living system; instead he was able to carry out limited measurements upon a single fluid, the blood. And the blood was treated by Magnus as a physical fluid, indistinguishable from other fluids, and certainly devoid of any "vital" qualities. This is quite a different problem from having to deal with an "innate heat" closely related to life itself.

As the theory of "innate heat" was replaced during the seventeenth century by the search for specific or limited causes for the heat phenomena of the body, the nature of

[28] Davy, "Experiments on the Blood," pp. 291–294, finds the same gases Magnus found and notes that he is now satisfied with these results "in spite of an opposite pre-existing bias" (p. 292).

[29] J. Davy was one of those who made explicit the implications of Magnus's work; *ibid.*, p. 298.

[30] Justus Liebig, *Animal Chemistry, or Chemistry in its Applications to Physiology and Pathology,* trans. William Gregory (2nd ed.; London, 1843), p. 35.

the research tools available also changed. The ancients had seen the general analogy between the "vital fire" in the heart and the flame of a lamp, but they never doubted that the internal fire arose from a unique source. The seventeenth-century chemist and physiologist insisted, on the basis of very little experimental evidence, that the processes involved in the generation of heat in flame and heat in the animal were identical. The tools he adopted to prove the identity were the tools of physics and chemistry. In this way the main features in the development of the theory of animal heat became tied to the evolution of new theories and new techniques in the physical sciences. The problem was biological only insofar as the research material happened to be a living organism. Parenthetically, we might remark that it is not surprising that many of the major advances in this study were made by students of chemistry and physics.[31]

As the techniques of physics and chemistry became more refined, and physical theories became subdivided, it is not surprising to see the study of animal heat reflect these changes. The eighteenth century saw the integration of theories of specific and latent heat in the explanation developed by Crawford. Lavoisier and Crawford were responsible for adopting the new chemistry of gases and combustion to the study of physiological heat. The real fragmentation of the problem came in the early nineteenth century. Although Magnus made a distinctive contribution to unraveling the mystery of where in the animal body heat was evolved, he made no mention in his paper of the question of the generation of heat. Similarly, the growing evidence

[31] It might also be pointed out that the chemist was as anxious to show that his new theory of, say, combustion had universal application and therefore was provoked into exploring the heat produced by animals.

that respiration took place in all the tissues of the body and not in the blood alone was overlooked by the students of animal heat.[32]

P. L. Dulong, although making direct calorimetric measurements of animal heat, provided a needed "justification" of the oxidation theory by redetermining the heats of combustion of hydrogen and carbon, thus giving ample evidence that the heat evolved in these processes was more than sufficient to account for the animal heat.[33] The extent to which the physiological explanation could still remain confused even though the physical "justification" had been provided is seen in the controversy which had earlier involved Dulong and his compatriot Despretz. In response to a prize contest announced by the Académie Royale des Sciences in 1821, both men had sought to determine "par des expériences précises quelles sont les causes, soit chimiques, soit physiologiques, de la chaleur animale."[34] The general method proposed was that of Crawford and Lavoi-

[32] Cruikshank, Abernethy, and Ingenhousz all recorded finding fixed air or carbonic acid emitted from the skin, as had Lavoisier and Séguin. None of them drew the connection to animal heat. William Cruikshank, *Experiments on the Insensible Perspiration of the Human Body, Shewing its Affinity to Respiration.* Published originally in 1779, and Now Republished with Additions and Corrections (London, 1795), pp. 61–91 and Advertisement. The work of Lazzare Spallanzani demonstrating tissue respiration is not even cited in most early studies of animal heat. Lazarus Spallanzani, *Memoirs on Respiration,* Edited from the Unpublished Manuscripts of the Author by John Senebier (London, 1804).

[33] Dulong had been a contestant, along with Despretz, for the 1823 Prix de l'Académie (see below). He withdrew his paper, however, because he did not feel confident in his results. He thence proceeded to the redetermination of the heats of combustion. See "Rapport fait à l'Académie des Sciences, sur un Mémoire de M. Dulong, Ayant pour Titre: De la Chaleur Animale, par MM. De Laplace, Chaussier, et Thenard, Rapporteur," *J. Physiol. Expt. Path.,* 3 (1823), 45–52; P. L. Dulong, "Mémoire sur la Chaleur Animale," *Ann. Chim. Phys.* [3], 1 (1841), 440–455; P. L. Dulong, "Recherches sur la Chaleur" [posthumously presented by J. D.], *Ann. Chim. Phys.* [3], 7 (1843), 180–183.

[34] "Programmes des Prix," *Ann. Chim. Phys.,* 19 (1821), 325–328.

sier, combustion analysis. The prize was awarded to Despretz in 1823 for his paper giving careful remeasurements of the heat produced by animals and the heat evolved in combustion. While he confirmed the theory that animal heat was a slow form of combustion, he did note a difference in the amount of heat produced by respiration and that produced during combustion. Despretz was unconcerned and chose to explain the small difference as being caused by movement of the blood and the friction of the different parts of the body on one another.[35] For the physicist, then, the mixture of physical causes raised no problem. The compatibility of these explanations with what really occurred in the animal did not seem important. An old, and discredited, explanation was brought in to account for that little heat not adequately explained by the more modern theory.[36] Only the later recomputation of the heats of combustion allowed these "extra" explanations to be quietly discarded.

The accurate determination of the temperatures of the various organs and tissues of the body removed many of the inaccurate guesses which had been at the base of different theories of animal heat. Becquerel and Breschet, who had adapted the thermocouple to the making of precise

[35] César Despretz, "Recherches Expérimentales sur les Causes de la Chaleur Animale," *Ann. Chim. Phys.*, 26 (1824), 352, 360. See also Paul S. Epstein, *Textbook of Thermodynamics* (New York, 1937), pp. 28–29.

[36] Curiously enough, when James Prescott Joule became involved in his priority dispute in thermodynamics he turned to a footnote he had appended to a paper in 1843 in which he reported a conversation with a physician who had "attempted to account for that part of animal heat which Crawford's theory had left unexplained, by the friction of the blood in veins and arteries." On the basis of another paragraph of amplification Joule hoped to strengthen his case for priority over Julius Robert Mayer. See J. P. Joule, "On the Calorific Effects of Magneto-Electricity, and on the Mechanical Value of Heat, Part II," *Phil. Mag.*, 23 (1843), 442.

temperature measurements, answered a long-contested question when they found that respiration actually did cool the blood and that blood in the pulmonary artery was perceptibly warmer than blood in the pulmonary vein.[37] The first clear indication that heat was actually generated in the tissues and not in the blood alone was provided by Hermann von Helmholtz when he measured the amount of heat produced by the twitch of isolated muscle.[38]

The theory of animal heat was no longer discussed as a whole, except in the textbooks. The working scientist dealt on the level of very limited physical-chemical phenomena, giving specific answers to specific questions. The parts, however, when tied together, whether they were temperature measurements or gas determinations, built a theory which had validity when viewed from outside biology. Tying their work closely to the events in chemistry and physics, the architects of the theory of animal heat clearly repudiated the idea that special biological laws governed the behavior of living systems.

The story of this fundamental shift in the nature of biological explanation has its beginnings in the establish-

[37] The following are among their more pertinent papers; Antoine César Becquerel and Gilbert Breschet, "Premier Mémoire sur la Chaleur Animale," *Ann. Chim. Phys.*, 59 (1835), 113–136; "(Second Mémoire sur la Chaleur Animale), Expériences sur Différens Cas Pathologiques," *Compt. Rend.*, 1 (1835), 28–30; "Recherches Expérimentales Physico-Physiologiques sur la Température des Tissus et des Liquides Animaux," *Compt. Rend.*, 3 (1836), 771–781; "Recherches sur la Chaleur Animale, au Moyen des Appareils Thermo-électriques," *Mus. Hist. Nat. Archives, Paris*, 1 (1839), 383–403; "Mémoires: 1⁰ Sur la Détermination de la Température des Tissus Organiques de Plusieurs Mammifères; 2⁰ Sur la Température Différente du Sang Artériel et du Sang Veineux," *Compt. Rend.*, 13 (1841), 791–797.

[38] Hermann Helmholtz, "Ueber die Wärmeentwickelung bei der Muskelaction," *Wissenschaftliche Abhandlungen*, vol. II (Leipzig, 1883), pp. 745–763. The paper was first published in 1848, but originally presented to the Physikalischen Gesellschaft in Berlin on 12 November 1847.

ment of the modern theory of animal heat. But only through the efforts of the many investigators who, throughout the nineteenth century, confirmed the physicochemical nature of many separate, basic physiological functions did biology achieve its belated revolution. Muscle no longer had its "special faculty" of contraction. Nerves lost their endowment of a special "vital" force destined to integrate the behavior of living things. Digestion and nutrition became the sources of the chemical energy which kept a biochemical system in action. Thus the locomotory, sensory, and vegetative functions of the living organism all became united in a single explanatory system understandable in the terms of modern physics and chemistry.

Selected Bibliography

Since full citation of secondary commentary and monographs has been included in the footnotes it seemed appropriate to present a bibliography of primary sources through which the reader would be enabled to scan the major sources utilized in this history of the theory of animal heat.

The following shortened titles have been used for the periodical literature: *Philosophical Transactions* in place of *Philosophical Transactions of the Royal Society*, London; *Observations sur la Physique* in place of *Introduction aux Observations sur la Physique, sur l'Histoire Naturelle et sur les Arts* (*Rozier's Journal*) and its successor *Journal de Physique, de Chymie, d'Histoire Naturelle* (*et des Arts*); *Annales de Chimie* in place of *Annales de Chimie, ou Recueil de Mémoire Concernant le Chimie et les Arts qui en Dépendent* (*et Spécialement la Pharmacie*) and its successor *Annales de Chimie et de Physique*; *Mémoires Académie Sciences* in place of *Histoire et Mémoire de l'Académie Royale des Sciences*, Paris; *Comptes-Rendus Académie Sciences* in place of *Comptes-Rendus Hebdomadaires des Séances, Académie Royale des Sciences*, Paris.

I. ANIMAL HEAT THROUGH THE SEVENTEENTH CENTURY

Akron, Philistion, Diokles von Karystos. *Die Fragmente der Sikelischen Arzte Akron, Philistion und des Diokles von Karystos.* Trans. Max Wellmann. Berlin, 1901.

"An Account of Two Books. 1. Tractatus Duo, Prior de Respiratione; alter de Rachitide, A. Joh. Mayow, Oxon, 1668," *Philosophical Transactions*, 3, No. 41 (1668), 833–835.

Aristotle. *Aristotle, Parva Naturalis, A Revised Text with Introduction and Commentary*. Sir David Ross. Oxford: Clarendon Press, 1955.

―――― *Generation of Animals*. Trans. A. L. Peck. Loeb Classical Library; Cambridge, Mass.: Harvard University Press; London: Heinemann, 1943.

―――― *De Iuventute et Senectute, De Vita et Morte, De Respiratione*. Trans. W. D. Ross. Oxford, 1908.

―――― *Parts of Animals*. Trans. A. L. Peck, Loeb Classical Library; Cambridge, Mass.: Harvard University Press; London: Heinemann, 1955.

―――― *De Sensu et Sensibili*. Trans. J. I. Beare. Oxford, 1908.

Avicenna. *A Treatise on the Canon of Medicine of Avicenna, Incorporating a Translation of the First Book*. O. Cameron Gruner. London, 1930.

Birch, Thomas. *The History of the Royal Society of London for Improving of Natural Knowledge, from its First Rise*. 4 vols. London, 1756–1757.

Borelli, Giovanni Alfonso. *De Motu Animalium*. 2nd ed. 2 vols. Lugduni Batavis, 1685.

Boyle, Robert. *The Works of the Honourable Robert Boyle*. Ed. Thomas Birch. 5 vols. London, 1744.

―――― *Tracts: Containing I. Suspicions about Some Hidden Qualities of the Air; with an Appendix Touching Celestial Magnets, and Some other Particulars*. London, 1674.

Charleton, Walter. *Physiologia Epicuro-Gassendo-Charltoniana: Or a Fabrick of Science Natural, Upon the Hypothesis of Atoms, Founded by Epicurus, Repaired by Petrus Gassendus, Augmented by Walter Charleton*. London, 1654.

Descartes, René. *The Philosophical Works of Descartes*. Trans. Elizabeth S. Haldane, G. R. T. Ross. 2 vols. Cambridge, 1931.

―――― *Oeuvres de Descartes*. Ed. Charles Adam and Paul Tannery. Vols. VI, XI. Paris, 1909.

Galen, Claudius. *On Anatomical Procedures*. Translation of the Surviving Books with Introduction and Notes by Charles Singer. London: Oxford University Press, 1956.

——— *Claudii Galeni Opera Omnia*. Ed. Carl Gottlob Kühn. Leipzig, 1821–1833.

——— *On the Natural Faculties*. Trans. A. J. Brock. Loeb Classical Library; Cambridge, Mass.: Harvard University Press; London: Heinemann, 1947.

Harris, John. *Lexicon Technicum: or, an Universal English Dictionary of Arts and Sciences: Explaining not only the Terms of Art, but the Arts Themselves*. London, 1704.

Harvey, William. *The Circulation of the Blood, Two Anatomical Essays by William Harvey together with nine letters written by him*. Trans. and ed. Kenneth J. Franklin. Oxford: Blackwell, 1958.

——— *Lectures on the Whole of Anatomy, An Annotated Translation of 'Prelectiones Anatomiae Universalis'*. C. D. O'Malley, F. N. L. Poynter, K. F. Russell. Berkeley, Calif.: University of California Press, 1961.

——— *De Motu Locali Animalium* (1627). Ed. and trans. Gweneth Whitteridge. Cambridge, Eng.: University Press, 1959.

——— *Movement of the Heart and Blood in Animals, an Anatomical Essay*. Trans. Kenneth J. Franklin. Oxford: Blackwell, 1957.

Helmont, Jean Baptista van. *Oriatrike or, Physick Refined, The Common Errors Therein Refuted, and the Whole Art Reformed and Rectified: Being a New Rise and Progress of Philosophy and Medicine*. Trans. J. C.   London, 1662.

Henshaw, Nathaniel. *Aero-Chalinos: Or, a Register for the Air; In Five Chapters*. 2nd ed. London, 1677.

Hippocrates. *Hippocrate, "L'ancienne médecine," Introduction, traduction et commentaire*. A.-J. Festugière. Paris: Klincksieck, 1948.

——— [Ancient Medicine]. *Philosophy and Medicine in Ancient Greece*. W. H. S. Jones. *Supplements to the Bulletin of the History of Medicine*, No. 8. Baltimore: Johns Hopkins Press, 1946.

Hippocratic Corpus. "*Peri kardies*. A Treatise on the Heart from the Hippocratic Corpus: Introduction and Translation." Frank R. Hurlbutt, Jr. *Bulletin of the History of Medicine*, 7 (1939), 1104–1113.

Hooke, Robert. *Micrographia* (1665). Ed. R. T. Gunther. In *Early Science in Oxford*, XIII. Oxford: University Press, 1938.

——— *The Posthumous Works of Robert Hooke, containing the*

Cutlerian Lectures and Other Discourses. Ed. Richard Waller. London, 1705.

Leonardo da Vinci. *The Notebooks of Leonardo da Vinci.* Ed., trans. and introd. Edward MacCurdy. 2 vols. New York: Reynal and Hitchcock, 1939.

Lower, Richard. *Tractatus de Corde item de Motu & Colore Sanguinis et Chyli in eum Transitu.* Trans. K. J. Franklin. *Early Science in Oxford,* IX. Ed. R. T. Gunther. Oxford, 1932.

Mayow, John. *Medico-Physical Works, Being a Translation of "Tractatus Quinque Medico-Physici* (1674)." Alembic Club Reprints, No. 17. 2nd. ed. Edinburgh: Alembic Club, 1957.

Nemesius. *Of the Nature of Man.* Trans. G. Wither. In *Cyril of Jerusalem and Nemesius of Emesa.* Ed. William Telfer. Library of Christian Classics, IV. Philadelphia: Westminster Press, 1955.

Plato. *Plato's Cosmology, The 'Timaeus' Translated with Commentary.* F. M. Cornford. New York, 1937.

Praxagoras of Cos. *The Fragments of Praxagoras of Cos and his School. Collected, Edited and Translated.* Fritz Steckerl. *Philosophia Antiqua,* VIII. Leiden: Brill, 1958.

Sylvius, Franciscus. *Opera medica.* Geneva, 1681.

Vesalius, Andreas. *De humani corporis fabrica.* Basel, 1543.

Willis, Thomas. *Dr. Willis's Practice of Physick, Being the Whole Works of that Renowned and Famous Physician.* London, 1684.

II. THE EIGHTEENTH CENTURY

Adet, P. A. *Réponse aux Réflexions sur la Doctrine du Phlogistique & de la Décomposition de l'Eau.* Philadelphia, 1797.

Arbuthnot, John. *An Essay Concerning the Effects of Air on Human Bodies.* London, 1733.

Beddoes, Thomas. *Chemical Experiments and Opinions Extracted From a Work Published in the Last Century.* Dedicatory note by T. Beddoes. Oxford, 1790.

—— *Observations on the Nature and Cure of Calculus, Sea Scurvy, Consumption, Catarrh and Fever: Together with Conjectures upon Several Other Subjects of Physiology and Pathology.* London, 1793.

Black, Joseph. "Experiments upon Magnesia Alba, Quicklime, and Some Other Alcaline Substances," *Essays and Observations,*

Physical and Literary. Read Before a Society in Edinburgh, II (1756), pp. 157–225.

—— *Lectures on the Elements of Chemistry.* Ed. John Robison. 2 vols. Edinburgh, 1803.

Blagden, Charles. "Experiments and Observations in an heated Room," *Philosophical Transactions,* 65 (1775), 111–123.

——"Further Experiments and Observations in an heated Room," *Philosophical Transactions,* 65 (1775), 484–494.

Boerhaave, Herman. *Elements of Chemistry: Being the Annual Lectures of Herman Boerhaave, M. D.* Trans. Timothy Dallowe. 2 vols. London, 1735.

—— *A New Method of Chemistry; Including the History, Theory, and Practice of the Art.* Trans. Peter Shaw. 3rd ed. corrected. 2 vols. London, 1753.

Braunio, J. A. "A Review of His *Dissertatio Physica Experimentalis, de Calore Animalium,*" *Medical Commentaries,* 1 (1773), 59–62.

Caverhill, John. *Experiments on the Cause of Heat in Living Animals and Velocity of the Nervous Fluid.* London, 1770.

—— *A Treatise on the Cause and Cure of the Gout.* London, 1769.

Cigna, Giovanni. "De Colore Sanguinis Experimenta Nonnulla," *Miscellanea Philosophico-mathematica Societatis Privatae Taurinensis,* 1 (1759), 68–74.

—— "Dissertation sur les Causes de l'Extinction de la Lumière d'une Bougie & de la Mort des Animaux Renfermés dans un Espace Plein d'Air," *Observations sur la Physique,* 2 (1772, 2nd ed. 1777), 84–105.

—— "De Respiratione," *Miscellanea Philosophico-mathematica Societatis Privatae Taurinensis,* 5 (1770–73), 109–161.

Cleghorn, William. *Disputatio Physica Inauguralis, Theoriam Ignis Complectens.* Edinburgh, 1779.

—— "William Cleghorn's 'De Igne' (1779)." Trans. and ed. Douglas McKie and Niels H. deV. Heathcote. *Annals of Science,* 14 (1958), 1–82.

Crawford, Adair. *An Experimental Enquiry into the Effects of Tonics, and Other Medicinal Substances, on the Cohesion of the Animal Fibre.* Ed. Alexander Crawford, M. D. London, 1816.

—— *Experiments and Observations on Animal Heat, and the Inflammation of Combustible Bodies. Being an Attempt to Resolve these Phaenomena into a General Law of Nature.* London, 1779; 2nd ed. with very large additions, London, 1788.

—— "Experiments and Observations on the Matter of Cancer, and on the Aerial Fluids Extricated from Animal Substances by Distillation and Putrefaction; Together with some Remarks on Sulphureous Hepatic Air," *Philosophical Transactions,* 80 (1790), 391–426.

—— "Experiments and Observations on the Stability of Heat in Animals." Reported to the Royal Society of London, December 6, 1787. Unpublished manuscript.

—— "Experiments on the Power that Animals, when Placed in Certain Circumstances, Possess of Producing Cold," *Philosophical Transactions,* 71 (1781), 479–491.

Crawford, John. *Cursus Medicinae; or a Complete Theory of Physic.* London, 1724.

Critical Review, or, Annals of Literature. "Review of *Experiments and Observations on Animal Heat . . .* by Adair Crawford," 48 (1779), 181–188.

—— "Review of *An Examination of Dr. Crawford's Theory of Heat and Combustion,* by William Morgan," 51 (1781), 212–216.

Cruikshank, William. *Experiments on the Insensible Perspiration of the Human Body, Shewing its Affinity to Respiration.* Published originally in 1779, and now republished with additions and corrections. London, 1795.

[Cullen, William.] *Institutions of Medicine. Part I. Physiology. For the Use of the Students in the University of Edinburgh.* Edinburgh, 1772.

Darwin, Erasmus. *Zoonomia; Or, the Laws of Organic Life.* Vol. I. Dublin, 1794.

Dobson, Matthew. "Experiments in an Heated Room. In a Letter to John Fothergill, M. D., F. R. S.," *Philosophical Transactions,* 65 (1775), 463–469.

Douglas, Robert. *An Essay Concerning the Generation of Heat in Animals.* London, 1747.

Elliott, John. *Philosophical Observations on the Senses of Vision and Hearing; to which are Added, A Treatise on Harmonic*

Sounds and an Essay on Combustion and Animal Heat. London, 1780.

Encyclopedia Britannica; or a Dictionary of Arts and Sciences, Compiled upon a New Plan . . . by a Society of Gentlemen in Scotland. 3 vols. Edinburgh, 1771.

An Enquiry into the General Effects of Heat; with Observations on the Theories of Mixture. London, 1770.

Fabre, P. *Réflexions sur la Chaleur Animale, pour Servir de Supplément à la Seconde Partie des Recherches sur Différens Points de Physiologie . . .* Paris, 1784.

Fordyce, George. "An Account of an Experiment on Heat," *Philosophical Transactions,* 77 (1787), 310–317.

Franklin, Benjamin. *Benjamin Franklin's Experiments; A New Edition of Benjamin Franklin's "Experiments and Observations on Electricity."* I. Bernard Cohen. Cambridge, Mass.: Harvard University Press, 1941.

—————— *Experiments and Observations on Electricity, Made at Philadelphia in America.* London, 1769.

Girtanner, Christoph. "Mémoires sur l'Irritabilité, Considerée comme Principe de Vie dans la Nature Organisée, *Second Mémoire*," *Observations sur la Physique,* 37 (1790), 139–154.

Goodwyn, Edmund. *The Connexion of Life with Respiration; or an Experimental Inquiry into the Effects of Submersion, Strangulation, and Several Kinds of Noxious Airs, on Living Animals.* London, 1788.

Hales, Stephen. *Statical Essays: Containing Haemastaticks; or an Account of Some Hydraulick and Hydrostatical Experiments Made on the Blood and Blood-Vessels of Animals . . .* Vol. II. London, 1733.

Haller, Albrecht von. *Dr. Albert Haller's Physiology; Being a Course of Lectures upon the Visceral Anatomy and Vital Oeconomy of Human Bodies.* 2 vols. London, 1754.

—————— *First Lines of Physiology.* Translated and printed under the inspection of William Cullen. 2 vols. Edinburgh, 1786.

Hassenfratz, Jean. "Mémoire sur la Combinaison de l'Oxigène avec le Carbone & l'Hydrogène du Sang, sur la Dissolution de l'Oxigène dans le Sang, & sur la Manière dont le Calorique se Dégage," *Annales de Chimie,* 9 (1791), 261–274.

Hewson, William. *An Experimental Inquiry into the Properties of the Blood. With Remarks on Some of its Morbid Appearances.* London, 1771.

Hey, William. *Observations on the Blood.* London, 1779.

Hunter, John. "Experiments on Animals and Vegetables, with Respect to the Power of Producing Heat," *Philosophical Transactions,* 65 (1775), 446–458.

————— "Of the Heat, &c. of Animals and Vegetables," *Philosophical Transactions,* 68 (1778), 7–49.

Irvine, William, and William Irvine, Jr. *Essays, Chiefly on Chemical Subjects.* Ed. William Irvine, Jr. London, 1805.

Irvine, William, Jr. "A Letter from Mr. Irvine Concerning the Late Dr. Irvine, of Glasgow, His Doctrine, which Ascribes the Disappearance of Heat, Without Increase of Temperature, to a Change of Capacity in Bodies, and that of Dr. Black, which Supposes Caloric to Become Latent by Chemical Combination with Bodies; with Particular Remarks on the Mistakes of Dr. Thompson, in His Accounts of these Doctrines," *Journal of Natural Philosophy, Chemistry, and the Arts* (Nicholson's), 6 (1803), 25–31.

Jackson, Seguin Henry. *Dermato-Pathologia; or Practical Observations, From Some New Thoughts on the Pathology and Proximate Cause of Diseases of the True Skin and its Emanations, the Rete Mucosum and Cuticle, with an Appendix Containing Further Observations on the Influence of the Perspirable Fluid in the Production of Animal Heat; and Remarks on the Late Theories of Scurvy; with the Particular View of Recommending the Oak Bark, as a New Marine Antiscorbutic; and as a Probable Antiseptic in Some other Putrescent Disorders.* London, 1792.

Lamarck, Jean Baptiste. *Recherches sur les Causes des Principaux Faits Physiques.* Vol. I. Paris, 1794.

Lavoisier, Antoine L. "De la Combinaison de la Matière du Feu avec les Fluides Evaporables, et de la Formation des Fluides élastiques aëriformes," *Mémoires Académie Sciences* (1777 [1780]), pp. 420–432.

————— "Expériences sur la Respiration des Animaux, et sur les Changements qui Arrivent à l'Air en Passant par leur Poumon," *Mémoires Académie Sciences* (1777 [1780]), pp. 185–194.

—— Mémoire sur la Combustion en Général," *Mémoires Académie Sciences* (1777 [1780]), pp. 592–600.

—— "Mémoire sur la Formation de l'Acide, Nommé Air Fixe ou Acide Crayeus, et que Je Désignerai Désormais sous le Nom d'Acide du Charbon," *Mémoires Académie Sciences* (1781 [1784]), pp. 448–467.

—— "Mémoire sur les Altérations qui Arrivent à l'Air dans Plusieurs Circonstances où se Trouvent les Hommes Réunis en Société," *Histoire [Mémoires] de la Société Royale de Médecine* (1782–83 [1787]), pp. 569–582.

—— *Traité Elémentaire de Chimie, Présenté dans un Ordre Nouveau, et d'après les Découvertes Modernes.* 3rd ed. 2 vols. Paris, 1801.

—— and Pierre S. Laplace. "Mémoire sur la Chaleur," *Mémoires Académie Sciences* (1780 [1784]), 355–408.

—— and Armand Séguin. "Second Mémoire sur la Respiration," *Annales de Chimie*, 91 (1814), 318–334.

—— "Second Mémoire sur la Transpiration des Animaux." *Traité Elémentaire de Chimie.* 3rd ed. Vol. II. Paris, 1801.

Leslie, P. Dugud. *A Philosophical Inquiry into the Cause of Animal Heat: With Incidental Observations on Several Phisiological and Chymical Questions, Connected with the Subject.* London, 1778.

Maclurg, James. *Tentamen Medicum Inaugurale, de Calore.* Edinburgh, 1770.

Magellan, Jean Hyacinthe. "Essai sur la Nouvelle Théorie du Feu Elémentaire, & de la Chaleur des Corps," *Observations sur la Physique*, 17 (1781), 375–386, 411–422.

—— "Extrait d'une Lettre de M. Magellan, de la Société Royale de Londres, sur les Montres nouvelles qui n'ont pas besoin d'être montées, sur celle de M. Mudge & sur l'Ouvrage de M. Crawford," *Observations sur la Physique*, 16 (1780 [1789]), 60–63.

—— "Lettre de M. Magellan à l'Auteur de ce Journal, sur le Mémoire Suivant," *Observations sur la Physique*, 17 (1781), 369–375.

Martine, George. *Essays Medical and Philosophical.* London, 1740.

—— *De Similibus Animalibus et Animalium Calore Libri Duo.* London, 1740.

—— "Some Thoughts Concerning the Production of Animal Heat, and the Divarications of the Vascular System, Being an Abstract from a Latin Treatise of the Heat of Animals; In a Letter to Dr. John Stevenson, Physician in Edinburgh," *Medical Essays and Observations*, 5th ed., vol. III (1771), pp. 111–129.

Medical Commentaries, "Medical News: Commentary on Animal Heat," 6 (1779), 98–103.

Menzies, Robert. *A Dissertation on Respiration*. Trans. with notes by Charles Sugrue. Edinburgh, 1796.

Moises, Hugh. *A Treatise on the Blood, or, General Arrangement of Many Important Facts, Relative to the Vital Fluid. With Cursory Observations on the Theory of Animal Heat. Interspersed with Pathological and Physiological Remarks from the Inductions of Modern Chemistry*. London, n. d. [1794].

Morgan, William. *An Examination of Dr. Crawford's Theory of Heat and Combustion*. London, 1781.

Mortimer, Cromwell. "A Letter to Martin Folkes, Esq.; President of the Royal Society, from Cromwell Mortimer, M. D. Secr. of the Same, concerning the natural heat of animals," *Philosophical Transactions*, 43 (1745), 473–480.

Moscati, Pietro. "Nouvelles Observations et Expériences sur le Sang et l'Origine de la Chaleur Animale," *Observations sur la Physique*, 11 (1778), 389–400.

Musgrave, Samuel. *Remarks on Dr. Boerhaave's Theory of the Attrition of the Blood in the Lungs*. London, 1759.

Nicholson, William. *A Dictionary of Chemistry*. 2 vols. London, 1795.

Peart, E. *The Generation of Animal Heat Investigated*. Gainsbrough, 1788.

Pitcairne, Archibald. *Elementa Medicinae Physico-Mathematica, libris duobus quorum prior theoriam, posterior praxim exhibet; In medicinae studiosorum gratiam delineata*. London, 1717.

—— *The Philosophical and Mathematical Elements of Physick*. Trans. John Quincy. 2nd ed. London, 1745.

Priestley, Joseph. *Experiments and Observations on Different Kinds of Air*. Vol. III. London, 1777.

—— "Observations on Respiration, and the Use of the Blood," *Philosophical Transactions*, 66 (1776), 226–248.

Rigby, Edward. *An Essay on the Theory of the Production of Animal Heat, and on its Application in the Treatment of Cutaneous Eruption, Inflammations, and Some other Diseases.* London, 1785.

Rolli, Paul. "An Extract by Mr. Paul Rolli, F. R. S. of an Italian Treatise, written by the Reverend Joseph Bianchini, a Prebend in the City of Verona; upon the Death of the Countess Cornelia Zangari & Bandi, of Cesena. To which are subjoined Accounts of the Death of Jo. Hitchell, who was burned to Death by Lightning; and of Grace Pett at Ipswich, whose Body was consumed to a Coal," *Philosophical Transactions,* 43 (1745), 447–465.

Rutherford, Daniel. "Daniel Rutherford's Inaugural Dissertation," trans. Crum Brown, communicated by Leonard Dobbin, *Journal of Chemical Education,* 12 (1935), 370–375.

Scheele, Carl. *Chemical Observations and Experiments on Air and Fire.* Introd. Torben Bergman, trans. J. R. Forster, notes Richard Krwan. London, 1780.

Séguin, Armand. "Observations Générales sur la Respiration et sur la Chaleur Animale," *Observations sur la Physique,* 37 (1790), 467–472.

―――― and Antoine L. Lavoisier. "Premier Mémoire sur la Respiration des Animaux," *Mémoires Académie Sciences* (1789 [1793]), pp. 566–584.

―――― "Premier Mémoire sur la Transpiration des Animaux," *Mémoires Académie Sciences* (1790 [1797]), pp. 601–612.

Senac, Jean Baptiste. *Traité de la Structure du Coeur, de son Action, et de ses Maladies.* 2nd ed. 2 vols. Paris, 1777 [1st ed. 1749].

Stevenson, John. "An Essay on the Cause of Animal Heat, and on Some of the Effects of Heat and Cold on Our Bodies," *Medical Essays and Observations,* 5th ed., Vol. V, pt. 2 (1771), pp. 326–413.

Whytt, Robert. *Treatise on the Vital and Other Involuntary Motions in Animals.* 2nd ed. Edinburgh, 1763.

Wilson, Patrick. "An Account of a Most Extraordinary Degree of Cold at Glasgow in January Last . . ." *Philosophical Transactions,* 70 (1780), 451–473; 71 (1781), 368–394.

III. THE NINETEENTH CENTURY

Allen, William, and William H. Pepys. "On the Changes Produced in Atmospheric Air, and Oxygen Gas, by Respiration," *Philosophical Transactions*, 98 (1808), 249–281.

———— "On Respiration," *Philosophical Transactions*, 99 (1809), 404–429.

Becquerel, Antoine C., and Gilbert Breschet. "Mémoires: 1ᵉ sur la Détermination de la Température des Tissus Organiques de Plusieurs Mammifères; 2ᵉ sur la Température Différente du Sang Artériel et du Sang Veineux," *Comptes-Rendus Académie Sciences*, 13 (1841), 791–797.

———— "Premier Mémoire sur la Chaleur Animale," *Annales de Chimie*, 59 (1835), 113–136.

———— "Recherches Expérimentales Physico-physiologiques sur la Température des Tissus et des Liquides Animaux," *Comptes-Rendus Académie Sciences*, 3 (1836), 771–781.

———— "Recherches sur la Chaleur Animale, au Moyen des Appareils Thermo-électriques," *Archives Museum d'Histoire Naturelle*, Paris, 1 (1839), pp. 383–403.

———— "(Second Mémoire sur la Chaleur Animale). Expériences sur Différens Cas Pathologiques," *Comptes-Rendus Académie Sciences*, 1 (1835), 28–30.

Berger, Jean F. "Faits Relatif à la Construction d'une Echelle des Degrés de la Chaleur Animale," *Mémoires de la Sociéte de Physique et Histoire Naturelle, Genève*, 6 (1833), 257–368; 7 (1834), 1–76.

Bernard, Claude. *Leçons sur la Chaleur Animale, sur les Effets de la Chaleur, et sur la Fièvre*. Paris, 1876.

Berthollet, C. L. "Sur les Changemens que la Respiration Produit dans l'Air," *Mémoires de Physique et de Chimie*, Société d'Arcueil, 2 (1808), 454–463.

Bessières, J. P. *Essai Historique et Critique sur la Chaleur Animale*. Toulouse, 1821.

Bostock, John. *An Essay on Respiration*. Parts I and II. Liverpool, London, 1804.

Brodie, Benjamin C. "The Croonian Lecture, on Some Physiological Researches, Respecting the Influence of the Brain on the Action of the Heart, and on the Generation of Animal Heat," *Philosophical Transactions*, 101 (1811), 36–48.

—— "Experiments and Observations on the Different Modes in which Death is Produced by Certain Vegetable Poisons," *Philosophical Transactions*, 100 (1811), 178–207.

—— "Further Experiments and Observations on the Action of Poisons on the Animal System," *Philosophical Transactions*, 102 (1812), 205–227.

—— "Further Experiments and Observations on the Influence of the Brain on the Generation of Animal Heat," *Philosophical Transactions*, 102 (1812), 378–393.

Broughton, S. D. "An Experimental Inquiry into the Physiological Effects of Oxygen and other Gases upon the Animal System," *Quarterly Journal of Science*, 29 (1830), 1–20.

Cabart. "Description de la Caisse du Calorimetre de M. Dulong," *Annales de Chimie* [3], 8 (1843), 183–189.

Carlisle, Anthony. "The Croonian Lecture on Muscular Motion," *Philosophical Transactions*, 95 (1805), 1–30.

Chossat, Charles. *Mémoire sur l'Influence du Système Nerveux sur la Chaleur Animale*. Paris, 1820.

Cornilleau, Auguste. *Essai sur la Chaleur Animale*. Collection Générale des Dissertations de la Faculté de Médecine de Strasbourg, XVI (1811), 16 pp.

Dalton, John. *A New System of Chemical Philosophy*. Part I. Manchester, 1808.

—— "On Respiration and Animal Heat," *Memoirs of the Manchester Literary and Philosophical Society* [2], 2 (1813), 15–44.

Davy, Humphry. *The Collected Works of Sir Humphry Davy*. Ed. John Davy. Vol. II. London, 1839.

—— *Researches, Chemical and Philosophical; Chiefly Concerning Nitrous Oxide, or Dephlogisticated Air, and its Respiration*. London, 1800.

Davy, John. "An Account of Some Experiments on Animal Heat," *Philosophical Transactions*, 104 (1814), 590–603.

—— "An Account of Some Experiments on the Blood in Connexion with the Theory of Respiration," *Philosophical Transactions* (1838), pp. 283–299.

—— "Miscellaneous Observations on Animal Heat," *Philosophical Transactions* (1844), pp. 57–64.

—— "Observations on Air Found in the Pleura, in a Case of Pneumato-thorax; with Experiments on the Absorption of Different Kinds of Air Introduced into the Pleura," *Philosophical Transactions*, 113 (1823), 496–516.

—— "Observations on the Temperature of Man and Other Animals," *Edinburgh Philosophical Journal*, 13 (1825), 300–311; 14 (1826), 38–46.

—— "On the Temperature of Man," *Philosophical Transactions* (1845), pp. 319–333.

—— "On the Temperature of Man within the Tropics," *Philosophical Transactions* (1850), pp. 437–466.

—— *Researches, Physiological and Anatomical.* 2 vols. London, 1839.

Despretz, César. "Lettre de M. Despretz à M. le Prof. De La Rive, Relative à l'Article de M. Dulong sur la Chaleur Animale," *Bibliothèque Universelle de Genève*, 34 (1841), 175–177.

—— "Recherches Expérimentales sur les Causes de la Chaleur Animale," *Annales de Chimie*, 26 (1824), 337–364.

Dulong, Pierre Louis. "Mémoire sur la Chaleur Animale," *Annales de Chimie* [3], 1 (1841), 440–455.

[——] "Rapport Fait à l'Académie des Sciences sur un Mémoire de M. Dulong, Ayant pour Titre: De la Chaleur Animale, par MM. De Laplace, Chaussier, et Thenard, Rapporteur," *Journal de Physiologie Expérimentale (et Pathologique)*, 3 (1823), 45–52.

[——] Presented by J. D. "Recherches sur la Chaleur," *Annales de Chimie* [3], 8 (1843), 180–183.

Edwards, W. F. "Animal Heat," *The Cyclopedia of Anatomy and Physiology.* Ed. Robert B. Todd. Vol. II. London, 1836–1839. Pp. 648–684.

—— "Sur l'Exhalation et l'Absorption de l'Azote dans la Respiration," *Journal de Physiologie Expérimentale (et Pathologique)*, 3 (1823), 19–28.

—— *On the Influence of Physical Agents on Life.* Trans. Hodgkin and Fisher. Philadelphia, 1838.

Ellis, Daniel. *An Inquiry into the Changes Induced on Atmospheric Air, by the Germination of Seeds, the Vegetation of Plants, and the Respiration of Animals.* Edinburgh, 1807.

Fourcroy, A. F. *A General System of Chemical Knowledge and its Application to the Phenomena of Nature and Art.* Trans. William Nicholson. 11 vols. London, 1804.

Gavarret, Jules. *Physique Médicale. De la Chaleur Produite par les Etres Vivants.* Paris, 1855.

Hale, Enoch, Jun. *Experiments on the Production of Animal Heat by Respiration.* An Inaugural Dissertation, Read and Defended at the Public Examination, Before the Rev. President and the Medical Professors of Harvard University, August 20, 1813. Boston, 1813.

Helmholtz, Hermann. *Wissenschaftliche Abhandlungen.* Vol. II. Leipzig, 1883.

Home, Everard. "The Croonian Lecture, on the Changes the Blood Undergoes in the Act of Coagulation," *Philosophical Transactions*, 108 (1818), 172–184.

Josse, F. *De la Chaleur Animale, et de ses Divers Rapports, d'Après une Explication Nouvelle de Phénomènes Calorifiques, avec l'Examen de l'Opinion de Différens Auteurs Modernes sur le Même Sujet.* Paris, 1801.

Joule, James Prescott. "On the Calorific Effects of Magneto-electricity, and on the Mechanical Value of Heat, Part 2," *Philosophical Magazine* [3], 23 (1843), 435–443.

Legallois, Julien J. C. "Deuxième Mémoire sur la Chaleur Animale," *Annales de Chimie* [2], 4 (1817), 5–23, 113–127.

Liebig, Justus. *Animal Chemistry, or Chemistry in its Application to Physiology and Pathology.* Trans. William Gregory. 2nd ed. London, 1843.

Magnus, Gustav. "Ueber die im Blute Enthaltenen Gase, Sauerstoff, Stickstoff und Kohlensäure," *Annalen der Physik und Chemie*, 40 (1847), 583–606.

Metcalfe, Samuel L. *Caloric, its Mechanical, Chemical and Vital Agencies in the Phenomena of Nature.* 2 vols. London, 1843.

Mons, J. B. von. *Sur l'Origine et sur la Distribution Uniforme de la Chaleur Animale.* Dissertation No. 118. Présentée et Soutenue à l'Ecole de Médecine de Paris, le 31 Août, 1808. Paris, 1808.

Müller, Johannes. *Elements of Physiology*. Trans. W. Baly. Vol. I. London, 1838.

Philip, A. P. Wilson. *An Experimental Inquiry into the Laws of the Vital Functions, with Some Observations on the Nature and Treatment of Internal Diseases*. Philadelphia, 1818.

Prout, William. "Observations on the Quantity of Carbonic Acid Gas Emitted from the Lungs During Respiration, at Different Times and Under Different Circumstances," *Annals of Philosophy*, 2 (1813), 328–343.

———— "On the Phenomena of Sanguification, and on the Blood in General," *Annals of Philosophy*, 13 (1819), 12–25, 265–279.

———— "Some Farther Observations on the Quantity of Carbonic Acid Gas Emitted From the Lungs in Respiration at Different Times and Under Different Circumstances," *Annals of Philosophy*, 4 (1814), 331–337.

Reid, John. "Respiration," *The Cyclopedia of Anatomy and Physiology*. Ed. Robert Todd. Vol. IV, pt. 1. London, 1847–1849. Pp. 325–368.

Spallanzani, Lazarus. *Memoirs on Respiration*. Edited from the unpublished manuscripts of the Author by John Senebier. London, 1804.

Spencer, Thomas. *Vital Chemistry. Lectures on Animal Heat*. Geneva, N. Y., 1845.

Thomson, Thomas. *A System of Chemistry*. 2nd ed. 4 vols. Edinburgh, 1804.

Tyndall, John. "Notes on Scientific History," *Philosophical Magazine* [4], 28 (1864), 25–51.

Welter, J. J. "Comparaison de la Quantité de Chaleur Dégagée par 1 Gramme d'Oxigène Brulant Divers Substances," *Annales de Chimie*, 19 (1821), 425–426.

———— "Observation sur la Quantité de Chaleur Dégagée Pendant la Combustion," *Annales de Chimie*, 27 (1824), 223.

Winn, J. M. "On a Remarkable Property of Arteries Considered as a Cause of Animal Heat," *Philosophical Magazine*, 14 (1839), 174–177.

Index

SUBJECT TO
LIBRARY RECALL

DATE DUE

GAYLORD   PRINTED IN U.S.A.